AMX30

MAIN BATTLE TANK

1960–2019 (AMX30B, AMX30B2 and derivatives)

First published in January 2020

A catalogue record for this book is available from the British Library.

ISBN 978 1 78521 648 0

Library of Congress control no. 2019943234

Published by Haynes Publishing,
Sparkford, Yeovil, Somerset BA22 7JJ, UK.
Tel: 01963 440635
Int. tel: +44 1963 440635
Website: www.haynes.com

Haynes North America Inc.,
859 Lawrence Drive, Newbury Park,
California 91320, USA.

Printed in Malaysia.

Senior Commissioning Editor: Jonathan Falconer
Copy editor: Michelle Tilling
Proof reader: Penny Housden
Indexer: Peter Nicholson
Page design: James Robertson

AMX30
MAIN BATTLE TANK

1960–2019 (AMX30B, AMX30B2 and derivatives)

Enthusiasts' Manual

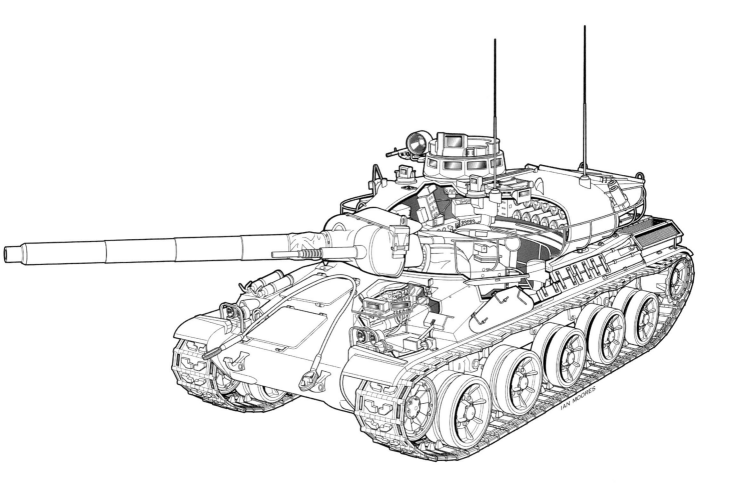

An insight into the development, construction and operation
of the AMX30 family of vehicles

M.P. Robinson and Thomas Seignon

CNE BOUVILLOIS

+61 0103

Contents

6 Foreword

Acknowledgements 7

8 Authors' preface

9 Introduction

12 Development of the AMX30

Production of the AMX30B 27
The AMX30B enters service 35
AMX32: the tank the French Army could
 not afford 41
Renewal: the AMX30B2 development, production
 and conversion programmes, 1980–91 51
AMX30B2 Brennus: the last upgrade 59

64 Anatomy of the AMX30B

Automotive drive train – engine/transmission/
 final drives, tracks and suspension 68
Communications 76
Water-crossing capability and doctrine 77
The NBC system 79

82 Main armament and sighting
 equipment

Secondary armament – 12.7mm and CN-20-F2 86
The Obus-G HEAT round 89
Sighting and optics – AMX30B 90
The AMX30B2's COTAC fire control system 95

98 Maintaining the AMX30

Servicing and non-technical inspections 100
Shared maintenance tasks 103
Driver 104

108 The AMX30 family of vehicles

AMX30D recovery vehicle 110
AMX30 Pluton 114
AU F1 self-propelled gun 119
AMX30 Roland and Shahine 132
AMX30-based engineers' vehicles 139
Mine-clearing vehicles 144

148 AMX30 exports

The AMX30 family in foreign service 150

162 The AMX30 series in combat

Service in the French Army 164
AMX30 ACRA (Arme Anti-Char RApide) 171
Postscript: the GIAT AMX40 172

176 Appendices

1 AMX30 characteristics 176
2 AMX30 units 178
3 Production chart of AMX30B2s 181
3 Production chart of AMX30B2s 184
3 Production chart of AMX30B2s 185

186 Glossary of terms and
 abbreviations

186 Further reading

186 Index

OPPOSITE *Capitaine Rouvillois*, an **AMX30B2 of the**
2e Régiment de Chasseurs. **The AMX30B and AMX30B2 were**
very popular with the majority of their crews. While the crew
positions were spartan in comparison to some other tanks,
an average-sized man soon found himself accustomed to its
layout. The basic AMX30B design was ergonomic, well armed
and easy to maintain. The improved AMX30B2 was an easier
tank to drive and incorporated an effective electronic fire
control system. All of the AMX30 series tanks included excellent
sights and vision devices for the crew. *(Charles Beaudouin)*

Foreword

Général de Division Charles Beaudouin

The AMX30 was a weapon system that served in France's first line of defence for nearly 40 years. This era encompassed generations of soldiers and included my father's career as an officer – and my own. My father, Jean Beaudouin, converted from the M47 to the new AMX30B at the head of the *2e Régiment de Cuirassiers'* 3rd squadron in 1972. He retired a colonel, having spent his working years as an officer of the *Arme Blindée Cavalerie*. I first got to know the AMX30B2 in the 1980s, commanding a platoon in the *2e Régiment de Chasseurs*. I went on to lead the 2nd squadron of AMX30B2s in the *503e Régiment de Chars de Combat*. It was the experience accumulated on the AMX30B2 with this squadron which prepared my men and myself for the honour and responsibility of validating and proving the Leclerc a year later.

It is striking how history crosses back upon itself. The AMX30 lived many incarnations. My first AMX30B2, 614 0103 *Capitaine Rouvillois*, had spent its first years as an AMX30B in my father's squadron in the *2e Régiment de Cuirassiers* 15 years earlier. On two occasions I even crossed paths with my father's old tank, 614 0134, by then also rebuilt into an AMX30B2. My father and I were lucky enough in our time to lead our respective squadrons for three years, a rare privilege in an army that staunchly observed 24-month commands. We shared a common passion and a common vocation – and we shared common accomplishments in the service of our country. There is a special pride in following in the footsteps (and in my case in the tracks) of one's father.

My father and I still recall the AMX30 together. It was quite a weapon, almost an

BELOW An AMX30B2 of the *4e Régiment de Dragons* serving in Operation Daguet in March 1991. The AMX30B2 proved an excellent combat vehicle when it saw battle, some 24 years into its service career.
(Thomas Seignon)

embodiment of everything the French Army wanted in a tank. The turret was remarkably well laid out, the TOP7 cupola gave an incomparable 360-degree panoramic view. The commander could easily survey the battlefield and the tank's surroundings without blind spots. The controls and optics were simply and intelligently placed. The armament was excellent for such a light tank, combining an accurate 105mm piece and a powerful 20mm coaxial automatic cannon. A progressive system of upgrades left only the armoured shell unchanged between the AMX30B of 1967 and the AMX30B2 Brennus of 1996. In this optimised form the AMX30 was fitted with the DIVT16 CASTOR thermal camera, a new Mack E9 engine and explosive reactive armour.

The 30-tonne tank also formed the basis for a very complete family of variants and achieved some notable export successes – and thus left its mark on the history of several armies. In France, its employment in the deserts of Kuwait in 1991 marked its swansong after nearly 25 years of tensely guarding France's border and its large sector in West Germany. By then the prospect of the Leclerc's 120mm gun, and digital architecture were already pointing the way ahead towards the future. And yet the AMX30 has survived in secondary roles to the present day, representing the last link with the kind of military technology that evolved directly from the experience of the Second World War. Much like the Leclerc that followed, the AMX30 bore the mark of France's design and doctrinal priorities. These machines place firepower, mobility and quality optics above the heaviest armour. The French battle tank designs are more comparable to quick and agile hunters than to the brute force of lumbering armoured mastodons.

A tank crew's collective spirit transcends the social classes and the different origins of its members. Each crewman is indispensable to the other and complementary to the shared mission – each plays a part in the survival of all on the battlefield. To all the crewmen I served with, my salute goes out with the greatest affection and with many fond memories. Today, a precious few AMX30B2s remain in our army, but I still take the occasional guilty pleasure of accepting a ride in one when I can (or even to

drive one myself). These few survivors reside at Sissonne. In the next year or two the iconic AMX30 will pass into history after a service that started in early 1967. My father and I, on behalf of those thousands of former crewmen, pay homage to the authors and to this book, which recounts the story of a remarkable weapon.

Général de Division Charles Beaudouin
Officier de la Légion d'Honneur, Commandeur de l'Ordre National du Mérite, Croix de la Valeur Militaire
July 2019

ABOVE This AMX30B2 Brennus is numbered 6904 0114, indicating it was converted from one of the last AMX30B2s newly built in 1990. It is one of the last operational AMX30s left in the French Army, nearly 53 years after the first AMX30B was delivered.
(Charles Beaudouin)

Acknowledgements

The authors humbly thank the following people: Simon Dunstan, Charles Beaudouin, Jerome Hadacek, Jean-Michel Boniface, Bernard Canonne, Pierre Delattre, Zurich 2RD, Peter Lau, Luis Pitarch Carrion, Carlos Antonio Arroyo Alonso, Olivier Julian, Jesús Pardo, Angel Ruiz, Pedro Miguel Paniagua, Olivier Carneau, Hugues Acker and especially the late Claude Dubarry. We also thank the staff of the CAAPC (Centre des Archives de l'Armement et du Personnel Civil) de Châtellerault, the Director and staff of the STAT (Section Technique de l'Armée de Terre), and the staff of the Musée des Blindées de Saumur.

We dedicate this book to Claude Dubarry and to all who designed, built and served on the AMX30 series of vehicles.

Authors' preface

The AMX30B entered service at the end of 1966 and was first used in combat by French forces in its updated AMX30B2 form in early 1991. Since late 1991 vast parks of AMX30Bs have been assembled at Gien, augmented since 1995 by AMX30B2s. These tanks served as a pool of spares for vehicles still retained in service, and as the post-Cold War era bloomed into the Afghanistan and Iraq Wars in 2001–3, these stored vehicles still supported substantial variant programmes. By 1995 the Leclerc started to replace the AMX30B2 comprehensively, a process largely completed in the *Arme Blindée Cavalerie* by 2006.

By this time the AMX30B2's greatest value was an opposing forces vehicle (*FORces ADverses*, or FORAD) available to the training establishments located at Mailly and Sissonne. As such they have endured to the present day in dwindling numbers. Today the Mailly FORAD has disappeared, leaving but a single platoon at Sissonne. A lone battery of AU F1TA self-propelled guns operated by the *40e Régiment d'Artillerie*, and a small number of EBGs (*Engin Blindé du Génie*) and mine-clearing tanks remain in service with the *13e Régiment du Génie*. The AMX30D recovery vehicle remains in service to support each.

At the time of writing, the AMX30B2 is in the full process of disappearing from French Army service, over five decades after the AMX30B was first shown in 1967. Its contribution to France's arms industry, its place among the iconic tanks developed since the Second World War and its impact on the many thousands of soldiers who crewed the AMX30 and its variants will never be forgotten in France. It has far outlived its contemporaries, the Chieftain, Leopard and M60A1, in the service of its army. Despite its adherence to a different gun and thinner armour, it remains one of the great tanks of the Cold War era. The AMX30B and AMX30B2 represented a golden age for the armaments industry in France, as did each variant adopted. The authors of this book truly hope to pay tribute to all who designed, built or crewed the AMX30 series AFVs.

Et Par Saint Georges, Vive la Cavalerie!
M.P. Robinson and Colonel Thomas Seignon
July 2019

RIGHT *Andelot*, an AMX30B of the *501e Regiment de Chars de Combat*, fitted with mine clearing rollers to prove the type's suitability as a mine clearing vehicle in 1990. The AMX30 was well regarded by the French Army for its ergonomic design, its firepower, its simplicity and as an adaptable basis for a range of derivative weapon systems, recovery and engineering vehicles. *(Collection STAT)*

Introduction

In France the study and construction of a 30-tonne medium tank design was undertaken by DEFA (*Direction des Etudes et Fabrication d'Armement*) in 1956. This decision followed a formal accord to collaborate on the FINABEL 3A5 common tank proposal signed by France, the Federal Republic of Germany and Italy. The desire to develop a light yet heavily armed medium tank came as a result of French experience in designing light and medium tanks stretching back to 1949. It also took note of Allied and German experience accumulated in the Second World War.

The French Army had not always wanted a lightweight medium tank. Before France was even completely liberated, in late 1944 DEFA was revived into prominence from pre-war obscurity as a minor government bureau into the state's central arms development organ. The first 'medium' tank (for it approached the size and weight of a Tiger) design it prepared was the *Char de Transition* ARL44. The unsuccessful ARL44 employed existing tank technology from French pre-war designs (in part derived from the Renault Char B1 (bis) design).

BELOW The AMX50 M4, a powerfully armed medium tank designed to achieve battlefield dominance through heavy firepower and sophisticated fire controls. Designed in 1948 and first built in 1949, it never achieved the level of mobility DEFA expected despite a development period that extended for nearly a decade (and which saw it redesigned as a heavy tank armed with a 120mm gun). *(Fonds Claude Dubarry, Collection du Musée des Blindées de Saumur)*

ABOVE The *Char Lorraine de 40 tonnes* (or CL 40t) of 1952. Its oscillating turret was a hallmark of French tank design in the 1947–56 period. The CL 40t was the first lightly armoured medium tank also equipped with a unique suspension system that included pneumatic rubber tyres, a feature which saved weight in comparison to the AMX50's suspension. Like the AMX50, it never entered production. *(Fonds Claude Dubarry, Collection du Musée des Blindées de Saumur)*

BELOW The American M47 was accepted gratefully by the French Army in 1952. Its arrival relegated a generation of French medium tank designs to obscurity despite the fact that the *Arme Blindée Cavalerie*'s leading thinkers were disappointed with its great weight, short range and old-fashioned design. *(Fonds Claude Dubarry, Collection du Musée des Blindées de Saumur)*

Shortly after the war ended, DEFA's AMX (*Ateliers d'Issy les Moulineaux*) bureau started work on an air-portable light tank and on large medium tank design at Satory. The smaller of these was developed into the AMX13, a successful design with the most powerful main armament yet carried by a tank of its class. The heavier of the two was the AMX50 M4, which also incorporated very modern features: an automatic loading system, a coincidence rangefinder and heavily sloped armour. The AMX50 was a medium tank along the lines of the Panther, and it too lacked a powerful enough engine to achieve the high mobility that the French Army demanded. The AMX50 design evolved over nine years, powered by optimised versions of the wartime German Maybach engines. Elements of its track and suspension system were derived directly from the Panther. The AMX50 was constructed in prototype form in 1949 and was seriously considered for production.

DEFA wished to produce the AMX50 to re-equip the French Army and to sell to NATO allies. Much like the contemporary AMX13, attempts were made to secure American Military Aid Plan funding. Prototypes of the AMX13 and AMX50 were shipped to the

USA and were evaluated by the US Army at Aberdeen Proving Ground in 1952. The surviving reports indicate that the Americans considered the AMX50 to be a promising design, but with a questionable powertrain. There were limits to American generosity, and funding could only be secured for the AMX13 light tank. DEFA proved incapable of resolving the issues with the Maybach engine, but the AMX50 development programme continued (as a heavy tank as its weight increased) until 1958.

The troubled development of the AMX50 encouraged private companies to present lighter medium tank designs to the army in the early 1950s. None of these resulted in success but two had technical merits worthy of discussion here. The first of these was the *Char Lorraine de 40 tonnes*, another large medium tank developed in 1952. It carried the same armament as the second AMX50 prototype but with much thinner armour, and in addition its unique suspension incorporated pneumatic tyres on the road wheels. At around the time that the Lorraine was evaluated, the M47 Medium Tank was obtained free of charge through the MAP programme to replace the M4 and M26 tanks that had equipped the

Arme Blindée Cavalerie since 1943 and 1950 respectively. The M47 entered service in 1954 and its availability rendered further development of a large medium tank unnecessary for the foreseeable future.

A much smaller medium tank design was developed from one of the failed contenders for the AMX13 design. The *Char Batignolles-Chatillon de 25 tonnes* of 1955 (known as the *Char BC*) was perhaps the most promising of the French tank designs of the 1950s. It carried a French 90mm gun – with firepower equivalent to the much heavier M47, designed in fact to fire American ammunition. When the lighter AMX13 Mle 58 prototype was demonstrated in 1956 armed with the new DEFA CN-105-57 105mm gun mounted in the FL12 turret, the *Char BC* ceased further development. Its major legacy was its powerful SOFAM petrol engine, but it proved that a 25-tonne medium tank carrying a 105mm gun could be built.

BELOW The *Char de Batignolles-Chatillon*, preserved today in this World of Tanks-inspired camouflage scheme in the Musée des Blindées de Saumur. This light medium tank was the first powered by the SOFAM petrol engine, used on the AMX30 prototypes and AMX30A *preseries*. *(Thomas Seignon)*

Chapter One

Development of the AMX30

Development of the AMX30 – from paper outline of specification FINABEL 3A5 into the finalised design set out between the two 'Chars de Definition' built in 1965 – took DEFA (and its successor DTAT) nine years. Development focused on three critical parameters, to which all other elements were subordinated: the D1512 gun, its associated optics, and the choice of a suitable engine.

OPPOSITE An AMX30A under test in cold weather conditions, probably during the winter of 1963/64. The AMX30 was expected to be capable of operations in the harshest of conditions, not only because of France's varied climes, but also in the hopes of securing the orders of north European allies. A rather discrete scorpion emblem is visible on the right side of the turret. *(Fonds Dubarry, Collection du Musée des Blindées de Saumur)*

By 1956 the French saw their future medium tank requirements in the 30-tonne range. The tank needed to be mobile, small in size and had to incorporate a nuclear, biological, chemical (NBC) filtration system to serve on a contaminated nuclear battlefield. It would dispense with the oscillating turrets characteristic of other French post-war tank designs since these were impossible to seal against radioactive contamination. A conventional three-man turret armed with a further development of the CN-105-57 gun and an optical rangefinder system was specified as a prerequisite for the new tank.

In the same period the French Army general staff was part of the FINBEL (France, Italy, Netherlands, Belgium, Luxembourg) staff committee within NATO. FINBEL was formed to counterbalance the numerous conferences conducted between the British, American and Canadian armies in the mid-1950s. The FINBEL liaison group admitted West Germany (and became FINABEL, 'A' being for Allemagne – Germany) when the latter was admitted to NATO in 1956. FINABEL permitted its members, all of whom depended on American MAP funding – to maintain some degree of consultative independence in military matters without recourse to American input. The French

representatives revealed DEFA's 30-tonne medium tank specification, which was endorsed by the other FINABEL representatives (including those from the newly formed *Bundeswehr*). As West Germany was authorised to rearm by NATO, and the *Bundeswehr* was looking beyond its unsatisfactory MAP-supplied weaponry, its representatives were soon involved in discussions with the French regarding a new medium tank. The 30-tonne tank specification of 1956 received the designation 'FINABEL 3A5', or more unofficially (and rather more memorably) as the *Europa Panzer*.

The FINABEL 3A5 specification became the basis for ongoing staff talks between the French, West German and Italian Armies, all then armed with the American M47. A bilateral agreement to adopt a common 30-tonne battle tank armed with a 105mm gun was signed by France and Germany on 27 October 1956. In May 1957 Italy formally joined the project. The French government was optimistic that DEFA could secure the role of senior partner in the 30-tonne project. The reason behind this was that DEFA was developing an excellent 105mm gun and was already producing AMX13s on four assembly lines for French and foreign orders. DEFA and the French government convinced themselves that French industry could expect to arm the FINABEL armies (and especially the *Bundeswehr*).

The French government's position ignored the fact that the West Germans had amassed a great deal of engineering and tactical experience regarding tank design and production during the Second World War. The Germans revealed little of their own plans, but they held a quiet conviction that they could manufacture a sophisticated medium tank for their own army, using their own heavy industry. In light of this unspoken divergence in national aims, the partnership was doomed from the start. Work began in earnest all the same. The decision to produce prototypes followed in 1958. Nationalist politics intruded into the Franco-German partnership after Charles de Gaulle became the prime minister and defence minister in June 1958. While de Gaulle could accept cooperation with West Germany, he was mistrustful of German intentions and insisted that France hold the role of senior partner in any collaborations.

BELOW A 1/10 scale wooden model produced at AMX in 1958, capturing the salient features of the **FINABEL 3A5** specification which evolved into the **AMX30.** By 1958 DEFA's designers realised that the development of heavier tanks like the AMX50 was not only too expensive, but was also fruitless given the technological progression of hollow-charge ammunition. *(Fonds Claude Dubarry, Collection du Musée des Blindées de Saumur)*

De Gaulle saw France as a great power – a country with its own strategic independence. He saw the development of a national independent nuclear capability as an absolute essential. The Colomb-Béchar Accord that promised to share nuclear technology between France and West Germany signed in 1956 had ignored his wishes and was revised at de Gaulle's behest in 1958. The West Germans naturally were deeply insulted by de Gaulle's policies and refused to accept the role of junior partner. By 1959 the West German economic recovery was in full swing and they were quietly looking beyond partnership with France. The existing works of scholarship from French sources are quite clear that West Germany's real position was not realised by the French until the first tank prototypes were already constructed. Thus, we should recognise that the development programme agreed in 1957 – of a common battle tank – was pursued with amiable cynicism by the West Germans and DEFA alike, for neither would have accepted anything other than their own tank!

The tripartite accord to develop a 30-tonne tank was premised (in theory) on the production of prototypes by France and West Germany, and an impartial evaluation of each. Production of the most suitable design would follow for all three armies. The basic French design held on to the SOFAM engine from the 25-tonne *Char de Batignolles-Chatillon*. The basic features common to all of the 3A5 proposals included a 105mm high-velocity gun, an NBC air filtration system, the capability of sealing the vehicle for submerged river crossing, the provision of a coincidence rangefinder to permit long-range target engagement and a high power-to-weight ratio. These qualities had never yet existed together in a production tank in the blend the French envisioned. The main characteristic ignored in comparison with British or American contemporary designs was in the thin armour protection visualised by DEFA and the West Germans. The tank would be armoured to a maximum of 80mm on the glacis and turret front, but on the sides and rear the armour was only sufficient to protect the vehicle from 20mm armour-piercing rounds. This choice came from a common belief that the day of heavy armour had passed with the development of hollow-charge munitions.

The decision to enter the prototype stage and advance to the comparative evaluation of the French and West German designs was agreed in 1960 by the West German defence

ABOVE The first AMX30 chassis prototype W510-207 is seen here during automotive trials in 1960. This prototype is easily identified by the large-size exhaust mufflers connected to the SOFAM petrol engine (which required large grilles on each side of the hull). The 'W' letter on the army licence plate identifies it as a prototype. *(Fonds Claude Dubarry, Collection du Musée des Blindées de Saumur)*

minister Strauss and *Ministre des Armées* Messmer on the French side. Two main parameters for comparing the French and West German tank designs were established at the same time. The first of these was a technical evaluation of each design. This included an extensive examination and testing of the turret systems, weapons and hulls by specialist personnel from each national trials establishment (and from each manufacturer). The second was a tactical evaluation to get the measure of each design in practical military terms. This stage employed crews from existing French and West German armoured units, who would each form a national platoon equipped with the AMX and West German prototypes.

The *Groupe de Travail Franco-Allemand A* (or FINABEL *Groupe A*) met at Kassel to decide the parameters for the common evaluation process on 25–26 February 1960 (and included Italian representation). A set of trials was agreed, establishing the criteria by which the French and West German prototypes could be evaluated by all three armies. The team assigned this work from *Groupe A* was simply designated the *Sous-Groupe de Travail*, and it met for the first time on 21–22 June 1960. Their meetings were spent studying how best to conduct tripartite tank trials. Joseph Molinié, in overall control of the French programme (and whose stewardship had already played its part in the success of the AMX13), later recalled:

The French and German engineers sat around the tables at dinner after a long day of meetings, melting the ice (as they say). We were all trying to get to know one another. I was seated next to an octogenarian German engineer who told us of his experiences developing the Tiger prototype during the war. As prototypes do, it broke down during its demonstration to the Führer. He still shuddered physically as he recounted the affair to us. …

Groupe A met again in Bonn on 19–20 December 1960 to fix the trial dates and to review progress made in the construction of prototypes. DEFA built its first two complete AMX30 prototypes in early 1961. These machines differed in detail from each other, but both were built to mount the SOFAM petrol engine. As a result, the machine featured a two-piece rear hull plate and a much smaller engine compartment than eventually seen on the production AMX30 hull. The first AMX hull prototype (W510-207) underwent automotive trials and was tested out in terms of general performance at Satory. These trials proved that the SOFAM petrol engine was a mature and suitable design.

Even turretless, the AMX hull looked decidedly modern, compact or even small in overall size. The hull glacis was well sloped, despite the relatively thin armour. The driver's position was on the left side of the glacis and featured a single-piece hatch incorporating three periscopes. Stowage compartments

BELOW The first prototype turret mounted on the first prototype chassis. The one-piece cast turret has a typical low profile, shaped like a water droplet (and possibly inspired by the Soviet T-55). The Vickers-type suspension (five main road wheels and four return rollers) with leading and trailing swing arms and torsion bars has already been selected. *(Fonds Claude Dubarry, Collection du Musée des Blindées de Saumur)*

were incorporated into the areas where the hull plates formed angles with the track guards. The rearmost lengths of the track guards mounted a prominent set of fabricated mufflers with large-diameter exhausts.

The original turret design consisted of a single-piece casting with a wide gun mantlet, cast bulges for the cross-turret rangefinder, and a dome-like appearance. The commander's cupola was a cast conical assembly incorporating seven periscopes and a rear-hinged one-piece hatch, reminiscent of the later types used on the Panther and Tiger tanks. The loader's hatch was hinged on its inboard edge, opening inwards towards the commander's cupola. The gunner and loader were each provided with a traversable periscope, and with a fixed episcope mounted in the turret wall. The turret bustle tapered into a pronounced bulge, the lower form sharply undercut. The second

ABOVE Prototype No 1, but with a second prototype turret. The shape of the upper part of the turret has been slightly redesigned to better integrate the optical coincidence rangefinder. *(Fonds Claude Dubarry, Collection du Musée des Blindées de Saumur)*

CENTRE Prototype No 1 is seen here in 1961 during trials on the Satory test area (near Versailles). This testing ground is still in use today, employed by Nexter Systems and Arquus. *(Fonds Claude Dubarry, Collection du Musée des Blindées de Saumur)*

RIGHT A head-on view highlights the conical commander's cupola, inspired by German wartime types. Although crew vision devices were not as obvious on the first two prototypes, the AMX30B that eventually resulted was better equipped with vision devices than many of its contemporaries. *(Fonds Claude Dubarry, Collection du Musée des Blindées de Saumur)*

ABOVE Prototype No 1 W510-207 during gunnery trials at Bourges. *(Fonds Claude Dubarry, Collection du Musée des Blindées de Saumur)*

prototype completed a few months later differed from the first in detail, but followed the same form.

The first AMX 30-tonne tank (weighing 34 tonnes) underwent preliminary trials from 20 to 25 February 1961 at Bourges, while the first *Standardpanzer* prototype underwent preliminary trials from 1 to 8 March at Meppen. A few weeks later, also at Meppen, the *Standardpanzer* and the first AMX30 prototype were demonstrated together. Gunnery tests proved that both guns were accurate. The *Standardpanzer*'s Daimler-Benz engine suffered from cooling problems and seized during the trial, leaving the German

tank broken down on the test ground, cutting short the proceedings. The AMX30 prototype performed very well on the other hand, much to the delight of the DEFA test crew, who parked their mount prominently in front of the mess hall as the trial broke for luncheon. On 10 April at Bonn, however, the French were dismayed when the Germans announced they intended to continue with the Daimler-Benz engine and the British gun. As far as the French were concerned, the German decision ignored the superiority of the French gun and ammunition and was politically motivated. The West Germans announced that their second prototype would be completed by the autumn of 1961 at the very earliest. A meeting was held at Satory on 15 December 1961 to finalise the comparative tactical and technical trial parameters for the AMX30A and West German prototypes on the basis of the work groups' recommendations. The second *Standardpanzer* prototype underwent

BELOW Prototype No 2 in 1962. The rear part of the chassis was significantly modified to fit a new and more powerful version of the SOFAM petrol engine. This also led to the modification of the mufflers, gaining their final form, and the addition of a fifth return roller to the suspension. From the French point of view, it was clear from the very start that in the balance of the priorities between firepower, mobility and protection, firepower was on the top of the list. *(Fonds Claude Dubarry, Collection du Musée des Blindées de Saumur)*

ABOVE The low profile, small size and general mobility of the vehicle were intended to compensate for the lower priority given to armour protection, which did not exceed 80mm on the frontal surfaces. *(Fonds Claude Dubarry, Collection du Musée des Blindées de Saumur)*

ABOVE Frontal view of AMX30 prototype No 2 in 1962. The general size of the first two AMX30 prototypes in comparison to the M47 was especially apparent in 1962 when this vehicle began trials. *(Fonds Claude Dubarry, Collection du Musée des Blindées de Saumur)*

BELOW Prototype No 2 was prepared for a series of technical tests at Satory in 1962. Trials focused on mobility, including slope climbing, trench crossing and deep wading. The gunnery trials followed at Bourges. *(Fonds Claude Dubarry, Collection du Musée des Blindées de Saumur)*

RIGHT Prototype No 2 in 1962 on a muddy track, typical of a military training ground – and probably seen at Mailly. At this stage of testing, the SOFAM petrol engine (a flat 12-cylinder petrol engine delivering 650hp at 2,750rpm) was still retained. *(Fonds Claude Dubarry, Collection du Musée des Blindées de Saumur)*

BELOW LEFT A front view of the AMX30A No 4 (234 0287). The mantlet infrared and white light projector changed in form and location before production. The vehicle looks considerably shorter without a TOP7 cupola fitted. The glacis is unencumbered with battery compartment covers and the driver's hatch features a single integral periscope. The cross turret rangefinder covers are in the open position. Frontal armour on the turret, gun mantlet and glacis was 50mm thick. *(Fonds Claude Dubarry, Collection du Musée des Blindées de Saumur)*

BELOW RIGHT Viewed from the rear we can see how the AMX30A's layout differed considerably from the production vehicle. The family resemblance to the AMX30B is already plain to see. *(Fonds Claude Dubarry, Collection du Musée des Blindées de Saumur)*

technical trials at Meppen from 24 to 28 September 1962. The second French prototype underwent its own trials at Mailly and Satory from 8 to 12 October 1962.

Seven AMX30A prototypes were constructed in 1962–63, all with detail differences in turret (then manufactured under contract by the *Compagnie des Ateliers et Forges de la Loire* or CAFL) and hull (assembled at AMX). The AMX30A turret had evolved considerably as ergonomic trials resulted in a longer, flatter-topped casting. The commander was provided with the SAMM (*Societé pour les Applications des Machines*

Motrices) S470 cupola then already in production for the AMX13 VTT armoured personnel carrier, mounting an externally served 12.7mm machine gun. The hull's form also evolved, enlarged in part to accommodate the diesel powerplant considered for production models, as well as to better accommodate the driver's position (which was distinguished by its redesigned hatch incorporating a single wide periscope).

The gun cradles fitted in the original seven AMX30 prototype turrets were of two different types. Three of the turrets had gun mountings with magnesium alloy cradles, while the other

ABOVE AMX30A No 4 at Mailly-le-Camp on 6 June 1963. This date was selected for the presentation of the new tank to the army. *(Fonds Claude Dubarry, Collection du Musée des Blindées de Saumur)*

BELOW Many of the features seen on AMX30A No 4 (234-0287) underwent considerable development prior to reaching the form seen on the production AMX30B. The suspension shock absorbers, road wheels, turret, driver's position and commander's cupola are the most obvious of these. *(Fonds Claude Dubarry, Collection du Musée des Blindées de Saumur)*

ABOVE This AMX30A was photographed while under evaluation on the Mourmelon training ground, in eastern France. After the tripartite evaluation in October 1963, a further trial phase took place to create standard operating procedures to develop platoon tactics. *(Fonds Claude Dubarry, Collection du Musée des Blindées de Saumur)*

BELOW An AMX30A being evaluated in suspension trials. The foremost road wheel station on the AMX30's suspension system incorporated a leading axle arm, attached to a shock absorber and sprung on a torsion bar. *(Fonds Claude Dubarry, Collection du Musée des Blindées de Saumur)*

four employed a heavier steel gun cradle design. While both types proved suitable to task, the stronger steel type was adopted for the AMX30B – despite a heavier weight. These seven vehicles were given official *numeros d'immatriculations* by the army in order to be operated in user trials. These were 234-0285, 234-0286, 234-0287, 234-0288, 234-0289, 234-0290 and 234-0291.

It had already been agreed during meetings held at Munsingen and at Stuttgart on 4–5 April 1962 that the platoon trials would take place. Discussions followed regarding the best possible testing ground on which to conduct the trials. The Munsingen training ground was well known for its firing ranges and was considered, but security factors precluded a tripartite evaluation on that base. The 'security arrangements' necessary were no doubt to keep the tripartite's Anglo-Saxon allies well clear of the proceedings, and one might presume any Soviet spies away as well! The French suggested the Mailly base, in Champagne. Mailly was long held as one of the best testing grounds and training areas in France, and it was located in reasonable proximity for the West Germans. The West Germans countered with the suggestion of conducting the trials at Bergen-Hohne. The final decision was deferred but the question was revisited when the work group next assembled at Rome on 7–8 May.

The work group's military committee finally chose Mailly as the site for the trials, authorising the inclusion of observers from the Belgian and Netherlands Armies (to the exclusion of all others). The resolution was taken to have a tripartite military committee meet within the month at Mailly to finalise a schedule for the tactical trials. Throughout the week of 23–30 May 1963 the details were worked out, fixing the date for the comparative platoon trials on the week of 16 September 1963. Single vehicle technical evaluation of the French tank would proceed from 10 to 16 October at Bourges and Satory, and of the West German tank at Meppen from 18 to 24 October. Reports from the French archives at Chatellerault written at the end of these trials still exist. These are paraphrased into a summary from notes made in part at Meppen on 24 and 25 October 1963 after the German tank's technical evaluation was complete.

The tactical phase of the trials was initially planned to encompass a period of nine or ten months (and to be conducted in all three countries). Without the use of the Munsingen training area, excluded in order to limit costs, this resulted in a shorter programme limited to a one-month period. For this trial, the West Germans had seven tanks available, the French had six, and each one of the trials were conducted on a five-tank platoon basis. The additional tanks were kept as reserves to permit full availability in case of mechanical failures.

Technical trials with reference vehicles were conducted by the national technical organisations (DEFA and Erprobungstelle *91). It must be noted that only the reference prototype of the AMX30 was equipped with the Hispano-Suiza HS-110 multi-fuel engine with the 5 SD 200 gearbox, unlike the examples employed in the tactical trials (which were all equipped with SOFAM petrol engines and the 5 S 200 gearbox).*

The trials were conducted under the overall direction of the Italian armoured corps' Colonel Fiandini. The three national delegations were led by Colonel Ing. Icken (West Germany), Colonel Delli Colli (Italy) and Lieutenant-Colonel Huberdeau

(France). Each delegation was composed of five armoured corps officers and military engineers. A Belgian and a Netherlands delegation, each composed of three officers, also observed throughout the trial.

For the tactical stage of the trials the 6 AMX30s were delivered to Mailly between 17 April and 4 June 1963 and these had an average of 2,200km on their odometers at the start of the trials. Each was equipped with the SOPELEM coincidence rangefinder, the SOFAM petrol engine and two vehicles were equipped with an improved gearbox. The seven West German tanks were practically brand new with an average of 600km on the odometers. They were equipped with rangefinders capable of stadiametric or coincidence functions, and one of them was equipped with a multi-fuel engine.

BELOW AMX30A No 4, 234-0287, at the Mailly trial in October 1963. The pre-production *Standardpanzer* **No 13 is visibly larger than the AMX30A. The production AMX30B weighed 11% less than the Leopard in combat order. We can see that the drivers sit on opposite sides, indicative on how the two 105mm guns were loaded and the crew layouts in the two vehicles. Interestingly enough, another key difference between these two tanks is the position of the rangefinder, provided to the commander in the French tank and to the gunner in the German design. This dictated a different position of the rangefinder openings in the two turrets.** (Fonds Claude Dubarry, Collection du Musée des Blindées de Saumur)

ABOVE AMX30A 234-0291 during cold-weather trials. The AMX30B production vehicle performed well in snow, but in French use the coldest weather they saw was a West German winter. In deep snow the AMX30's relatively low weight, good power-to-weight ratio and ground clearance were an advantage. Of course the vehicle's ability to climb in extreme icy conditions was as limited as any tracked vehicle, but the AMX30's tracks were well designed, adhering effectively and giving little trouble. Because France shared alpine frontier regions with Italy and Switzerland (and a mountainous border area with Spain), proving the new tank in the harshest conditions was an absolute necessity. *(Fonds Claude Dubarry, Collection du Musée des Blindées de Saumur)*

RIGHT Five of the AMX30As (including 234-0288 seen here), were employed in an experimental platoon as part of the M47-equipped *501e Régiment de Chars de Combat* in the months following the Mailly tripartite trial. *(Fonds Claude Dubarry, Collection du Musée des Blindées de Saumur)*

With the exception of the comparisons focused on the radios and the NBC systems, the bulk of the comparative trials were conducted to evaluate the mobility of the two tanks, and their armament and fire controls. During endurance testing two French and three West German tanks were eliminated due to breakdowns. On the whole, the West German chassis demonstrated slightly better performance, as one might imagine with its rather larger hull size. On the other hand, in terms of armament performance, the trials demonstrated very much in favour of the AMX30, especially at ranges in excess of 1,500m.

For the technical evaluation, a single reference vehicle was employed from each side. The West German reference vehicle had to be replaced after defects appeared in the turret systems following a test firing. The effectiveness of the French D1512 gun when firing OCC Mle 61 [sic] ammunition was demonstrated. The complete penetration of all test plates (170mm inclined to 50 degrees and 120mm inclined to 60 degrees and 150mm vertical plates) was achieved with the Obus-G round at all range distances. The West German tank, employing the 105mm L7 gun firing APDS ammunition, penetrated only 25% of these targets, and while this total improved with American M456 HEAT ammunition, it never approached the results achieved with the OCC Mle 61 [sic].

RIGHT Given the large audience in the background, the wading demonstration seen here must have taken place after the Mailly trial of October 1963, very possibly at Satory. The AMX30A seen here (234-0289) is dripping water after wading to turret depth. *(Fonds Claude Dubarry, Collection du Musée des Blindées de Saumur)*

CENTRE The same tank, seen from the rear. The basic waterproofing equipment that was standard for the AMX30A included rubber seals on all hatches, a turret ring seal and air aspiration through the turret. The exhaust system included flaps to prevent water ingress. *(Fonds Claude Dubarry, Collection du Musée des Blindées de Saumur)*

BELOW AMX30A prototype No 4, *immatriculation* 234-0287. Note how the road wheels equipping this vehicle have been changed to a simple dish pattern. The production AMX30B's road wheels were slightly different again, incorporating eight reinforcing ribs. *(Fonds Claude Dubarry, Collection du Musée des Blindées de Saumur)*

RIGHT A photograph of AMX30A prototype No 7. Note the stylised scorpion crest applied in positive and negative stencils on the turret side. The turret mounts no fewer than three wireless antennae, and the vehicle is stowed for operations. The turret bustle on the AMX30A lacked the stowage bin seen on production vehicles. *(Fonds Claude Dubarry, Collection du Musée des Blindées de Saumur)*

RIGHT The difference in terms of size and silhouette between AMX30A No 5 and the much heavier American M60A1. Logically enough, the design differences between these two tanks dictated the armour protection, weight and fuel consumption characteristics of each. *(Fonds Claude Dubarry, Collection du Musée des Blindées de Saumur)*

RIGHT This AMX30A weighed approximately 16 tonnes less than an M60A1, for an equal level of firepower. The difference in height, width and even the width of the tracks is evident. *(Fonds Claude Dubarry, Collection du Musée des Blindées de Saumur)*

At the end of these trials, which were among the first of their kind to be conducted in NATO, Général Doin (the head of the *Section Technique des Armées*) took stock of the two medium tank designs. On 23 November 1963 he wrote to Pierre Messmer, the *Ministre des Armées*, explaining his overall impression of the trial: 'If the superiority of the AMX30 in terms of its armament was proven without contest in these trials, we should be more prudent on the matter of the chassis. The West German tank proved at least an equal in this parameter to its French contemporary. …'

Production of the AMX30B

DEFA was reorganised as DTAT (*Direction Technique des Armements Terrestres*) in 1964 in order to render the national armaments complex more efficient and to close obsolete facilities. In 1972 DTAT was further rationalised into GIAT (*Groupement des Industriels de l'Armement Terrestre*). The AMX30B production programme represented a massive effort spread over the state-owned arsenals pursued between 1965 and 1981. Overall responsibility for the design through its life cycle was retained by the AMX bureau at Satory, but the AMX workshops only assembled the AMX30 prototypes. Responsibility for overall production was given to the Arsenal de Roanne (known hereafter by its acronym ARE), the largest armoured vehicle manufacturing facility in France. The responsibility for the manufacture of the AMX30's largest sub-assemblies were shared between state-owned facilities with a long history of specialisation.

With the Franco-German trials complete at the end of 1963 and both nations satisfied with their own product, no further pretence was observed about adopting a common medium tank. The West Germans had by then committed substantial resources to producing the *Standardpanzer*. The AMX30 had already also been voted to be ordered into production in July 1963 in the national assembly, well ahead of the Mailly trials. The final construction pattern for the AMX30 was set on 19 November 1963, on the basis of AMX30A hull prototype No 3 and AMX30A turret prototype No 4.

ABOVE One of the two *chars de définition* (or *avant serie*) built in 1965, serial 254-0745. This particular tank was the first AMX30 fitted with the TOP7 cupola. This tank featured in many of the photos used in the creation of the original AMX30B training manuals, although it differed in several areas from a production AMX30B. *(Fonds Claude Dubarry, Collection du Musée des Blindées de Saumur)*

The army had made its final choice on the new tank's form before Général Doin's letter describing his impression of the Mailly trial was even written. The AMX30 was theoretically ready for production (as the AMX30B) at the end of 1963, and it was estimated at the time that a maximum of 160 vehicles could be completed per year from the single ARE assembly line.

The harsh reality was that due to budget shortfalls in 1964, series production of the AMX30 was delayed. France's weak economic position in 1963–64 came as a result of the costly withdrawal from Algeria, and the pressing need to modernise the French Air Force before the Army. In 1964 debates raged about how the new tank's related variants could be prioritised. There was every intention of supplementing the AMX30B with the AMX30 ACRA guided missile tank, planned to enter production in 1968. Production of a recovery variant, specialised engineers' variants and artillery vehicles based on the AMX30 chassis were all anticipated.

In the months between the first order in late 1963 and production of the AMX30B in 1966, much work was completed at DTAT to design an entire family of vehicles based on the AMX30 chassis. The AMX30As were

ABOVE **Most obviously, 254-0745 was equipped with the PH8A IR/white light projector, which was replaced with the much lighter PH8B on production vehicles in order to save weight. It also tested out the concept of carrying the battery compartment covers side by side – a pattern adopted later for the AU F1 self-propelled gun. The layout of the radio antenna bases also still followed the pattern seen on the AMX30A.** *(Fonds Claude Dubarry, Collection du Musée des Blindées de Saumur)*

BELOW **Both tanks were tested extensively by the** *Section Technique des Armées***, and by the DGA. They were also used to train select crewmen from the 503e RCC during the course of 1966. Their subsequent fate is unknown.**
(Fonds Claude Dubarry, Collection du Musée des Blindées de Saumur)

employed to test out some available options – for example, 234-0097 was fitted with a SAMM S401 30mm Bitube DCA turret. This exercise proved the need for a dedicated auxiliary engine for any weapon system that employed complex electronics on the AMX30 chassis. Some 30 months passed before the first AMX30B was produced.

The whole of 1964 was spent tooling up the production facilities at Bourges, at Tarbes, Limoges and at ARE. The tooling for the production of hulls, suspension components and gearboxes was installed in the Somme and Marne complexes at ARE.

In 1965 two *chars de définition* (definition vehicles) were manufactured by the various DTAT production centres under the direction of the ARE. These were delivered to the army in September. One mounted the first TOP7 cupola, which was developed at Saint-Étienne. Contemporary documents describing the AMX30A from 1963 noted that a cupola armed with a '20mm automatic cannon' would be standard for production vehicles. This proved impossible to accomplish given the size of the TOP7 (*Tourelleau d'Observation Panoramique No 7*), and in fact the decision was subsequently taken to arm the production cupola with a 7.62mm machine gun and to provide a 12.7mm coaxial weapon (as fitted to both of the 1965 *chars de définition* pre-

production tanks). One of the AMX30 *de définition* was sent to the *Section Technique de l'Armée* (STA) for evaluation. Upon its return to the manufacturer both pre-production tanks were tested exhaustively at Mailly, Satory and on the ranges at Bourges. These trials lasted from 30 November 1965 until June 1966.

The *chars de définition* served to finalise the production AMX30B's features. These included the relocation of the turret's exterior stowage and the battery sealing plates on the glacis, changes in design to the infrared searchlight

and its mounting and of course the validation of the new TOP7 cupola. They finally served to train the first AMX30 drivers – and in early 1966 the first 40 AMX30Bs were cleared for production at ARE. The first two of these were delivered to the army in June 1966 but production began so slowly that prototype vehicles continued to be employed to train complete crews from the *503e Régiment de Chars de Combat* (503e RCC), earmarked as the first AMX30 regiment. The AMX30B design was set at the beginning of 1966. The 105mm

CN-105-F1 gun was manufactured at the *Arsenal de Bourges* (later known as the EFAB, France's most prominent artillery design and production centre), the optics and cupola were designed, assembled and integrated at the AMX/APX (*Atelier de Puteaux*) facility and under contract to private firms.

The impact of over two years of delay on

DTAT's ability to market the AMX30B in the face of the Leopard (which was being produced on four assembly lines) certainly affected international sales in the 1960s. This was especially true within the FINABEL countries – none of whom had joined France in adopting the AMX30. The annual production figures for the AMX30 and its variants have never been released, and (as with much that surrounds French Cold War armaments plans) these remain secret. The first order for 900 AMX30Bs was funded in the 1964–70 armaments programme, with the first 400 vehicles being ordered for delivery before 1970. Records indicate that by the end of 1969 the French Army had received 344 AMX30Bs and 10 AMX30D recovery vehicles, sufficient to equip two armoured divisions. Approximately 300 more were delivered in the period 1971–74. Between 1975 and 1981 the number of AMX30Bs delivered to the French Army annually slowed considerably – around 270 vehicles in all (with a further production to fill export orders and the French orders for artillery, missile launcher and recovery variants). The last of these were delivered in 1982, from an order for 60 French AMX30Bs outstanding in 1981. Annual production was normally sufficient to equip two regiments per year.

Foreign orders completed from the single ARE production line delayed the manufacture of variants, but by the end of 1970 the national arsenal production system was functioning extremely well. The ARE complex by then had enough capacity to service significant export orders. A national assembly committee delegated to audit the AMX30 programme reported the government's intentions for the production programme and its progress up to May 1974.

BELOW An AMX30B hull mounted on a massive jig at ARE (Arsenal de Roanne). The front of the hull was formed from multiple castings and plate sections welded together. A production target of 18 tanks a month at ARE was originally envisaged; however, in 1966–68 production of only 10, and then 13 tanks a month was achievable on the single production line, which was to produce the initial French order of 900 battle tanks. *(Fonds Claude Dubarry, Collection du Musée des Blindées de Saumur)*

LEFT The rear end of the hull was equally complex due to the multiple armour plate sections. No fewer than 150 different plates and castings were welded together to build an AMX30B hull. In 1969 an annual maximum of 156 AMX30Bs could be managed, which permitted the re-equipment of two and a half regiments per year, and virtually no spare capacity for foreign orders. *(Fonds Claude Dubarry, Collection du Musée des Blindées de Saumur)*

LEFT A colour photo taken on the ARE assembly line shows partly completed hulls, one of which is in the process of undergoing a powerpack installation. The complete hulls were integrated with their power trains at ARE. *(Fonds Claude Dubarry, Collection du Musée des Blindées de Saumur)*

LEFT The turret was cast and appointed with all of its weapons and equipment from the EFAB (Etablissement de Fabrication d'Armes de Bourges) and APX (Atelier de Puteaux) in the Atelier de Tarbes (ATS). *(Fonds Claude Dubarry, Collection du Musée des Blindées de Saumur)*

By then the AMX30 ACRA plan had been cancelled and the AMX30D was in production. It reported that 1,021 AMX30Bs had up to then been ordered (by the augmentation of the 1963 order with a further 121 vehicles). Production of the AMX30B was expected to be completed in 1976, and a further 40 were considered for a second French Army order. It reported how the monthly production at the ARE had passed from 10 in 1967, to 13 in 1968, 16 in 1970, 18 in 1972 and 20 in 1973. By the end of 1971, 754 AMX30Bs from the 1963 order had been delivered. By 1974 it was expected that 774 variants had been or would shortly be ordered – 59 AMX30H *Poseur de Pont*, 95 AMX30D recovery vehicles, 345 155mm GCT self-propelled guns, 46 AMX30P Pluton erector-launchers, and 214 AMX30R Roland surface-to-air (SAM) systems.

In addition, in May 1974 another 720 other gun tanks were expected to be produced for the French Army, starting in 1978. These were expected be based on the AMX30 but would be improved to include a 120mm main armament, automatic fire controls and a coaxial 20mm secondary armament (in effect, AMX32s). Some 425 of the AMX30Bs from the 1963 order were expected to be upgraded with automatic fire controls and the 20mm coaxial weapon at around the same time. By 1974, the individual

AMX30B unit cost had substantially decreased as production efficiencies were found within the GIAT organisation (or perhaps we should say the unit cost only increased by 22% at a time when inflation drove prices up some 58% across the French economy). The possibility of finding further economies by subcontracting in the production process was also recommended by the same committee.

Of course, the expected expansion of AMX30 production never came to pass. The bridge-layers were cancelled in 1975, the AMX32 was never taken up and smaller numbers of the AMX30R,

AU F1 and AMX30P were procured than had been optimistically expected in May 1974. The production programme remained largely that of 1970 right until the last new AMX30B2 was built in 1987. The AMX30B upgrade programme, which matured as the AMX30B2, was all that army budgets allowed in the second half of the 1970s. In the 1980s ARE was tasked with the production of variants of relatively small production batches of the AMX30B for export customers and the AMX30B2 for the French Army. It also directed the remanufacture of the large number of modernised AMX30 series vehicles, a task it shared in the 1980s with the large army workshops at Gien. Production at ARE reached its zenith between 1966 and 1986, the site being by then one of the most important manufacturing sites for AFVs in the Western world.

LEFT The AMX30B production line at ARE was complemented by a rebuilding line in the same facility and another at Gien. The biggest delays in producing enough AMX30Bs in the late 1960s were caused by the finite capacity available in DTAT's arsenals, which were simultaneously occupied with other production contracts. *(Fonds Claude Dubarry, Collection du Musée des Blindées de Saumur)*

RIGHT The Arsenal de Roanne included its own test track area, where this AMX30B is being put through its paces. The single production line at ARE proved adequate for the needs of the French Army. Had substantial orders materialised from the FINABEL countries, it is likely that this could have been addressed with production licences. *(Fonds Claude Dubarry, Collection du Musée des Blindées de Saumur)*

The AMX30B enters service

Production of course delayed the equipment of the army with the new tank. In 1966 Colonel Huberdeau, who had commanded the French delegation at Mailly, was instrumental in getting the new tank into service with the *Arme Blindée Cavalerie*. A veteran of the armoured battles of 1940 and of the liberation of France and Western Europe in 1944–45, he had later seen action in Algeria. He was given the honour of commanding the 503e RCC at the beginning of 1966, the first regiment selected to be equipped with the new tank. Highly respected by his men, Huberdeau drove the

ABOVE The first regiment equipped with the AMX30B was the *503e Régiment de Chars de Combat*. While this shows the Bastille Day parade of 1968, the same regiment paraded with two squadrons in the previous year, the occasion of the AMX30B's first introduction to the French public. The regiment had only 26 tanks available for the occasion. *(Bernard Canonne)*

unit's conversion from the M47 to the new tank with enthusiasm and was an inspiration to those who served under him. The men of the regiment were largely made up of young conscripts with a core of professional NCOs and experienced officers to train them into competent crewmen in time for the Bastille Day parade of 14 July 1967.

The first public appearance of the AMX30B came at the Satory arms display in June 1967

ABOVE An early production AMX30B climbing a steep incline during trials. This vehicle is in unit service and has received the name *Herbsheim* on the right side of the turret beneath the cupola. The first 40 AMX30Bs were manufactured between early 1966 and the middle of 1967. *(Fonds Claude Dubarry, Collection du Musée des Blindées de Saumur)*

alongside some 400 other French weapons. A spectacular display of the AMX30B's capabilities was given for the assembled French press corps at the ARE and again at the Satory test ground in the same month. The 503e RCC's crews trained hard to master the new tank,

which was a completely different animal from the familiar M47.

The AMX30B instantly became a symbol of France's technological prowess when two squadrons of the 503e RCC were included in the aforementioned Bastille Day parade in 1967. The parade attracted the attention of the national and foreign press, but the reality was that only 26 tanks were yet available for the regiment. The 503e RCC only received its full complement of 54 tanks in the subsequent months. The following year the regiment deployed all 54 of its tanks in the parade, a tradition observed annually for decades that followed.

Production in 1967–69 permitted the army to re-equip several regiments, normally a single squadron at a time. Re-equipment of units fielding the M47 in the *Forces Françaises en Allemagne* in West Germany was given priority, permitting the return of these heavier tanks supplied under the Military Aid Plan to American custody. A small number of cavalry regiments retained the AMX13 until enough AMX30Bs could be procured to permit full replacement, a process they completed in 1983. In French service the AMX30B was always simply known as the AMX30, the X30 or simply '*le trente*'.

The men of the *Arme Blindée Cavalerie* were very impressed with the AMX30B, which compared favourably to the fuel-hungry, heavy M47. Praise was heaped on the ergonomic

RIGHT The AMX30Bs' drivers were carefully supported during the conversion from M47s, and the process of getting the new tank into service passed quickly. The crews were very enthusiastic about the new tank, with the exception of drivers who had previously served on the M47! *(Fonds Claude Dubarry, Collection du Musée des Blindées de Saumur)*

LEFT Early production AMX30Bs seen at the southern fire range on the Mailly training ground. The absence of markings suggests recent deliveries, confirmed by the presence of M47s in the background. In the late 1960s the regiments were equipped one full squadron at a time, with the goal of ensuring that a maximum number of regiments were included, so mixed M47 and AMX30B inventories at regimental level could be expected in the first year. *(Fonds Claude Dubarry, Collection du Musée des Blindées de Saumur)*

turret design, the high rate of fire and accuracy of the gun, and on the AMX30's high road and cross-country speed in the hands of a seasoned driver. As France's tankers got used to their new mount, they eventually discovered its faults – just like any other tank of its time. In mechanical terms, the AMX30's greatest weakness was its manual BV5 SD transmission, and more specifically its Gravina clutch. The gearbox, especially when worn, presented well-known problems. These extended in extreme conditions to skipping the third gear, and a driver operating an AMX30B with a worn transmission might even have to stop the tank to change gear. This was compensated for in some measure by careful monitoring and inspection, and eventually by the improvement of individual components. Issues with the clutch were aggravated by inexperienced drivers, and by limited operational training hours due to fuel restrictions. The suspension design, with leading torsion bars on the first, third and fifth bogies, provided a rougher ride than a suspension of trailing torsion bars might have. The provision of an additional road wheel on each side might also have reduced the higher stress loading to the leading torsion bars. In service this occasionally resulted in broken torsion bars and tensioning units – certainly regularly enough to merit improvement.

RIGHT An AMX30B of the *11e Régiment de Cuirassiers* (11RC) in front of the drivers' training centre at Carpiagne, southern France. This training regiment was charged with drivers' and gunners' training for the *Arme Blindée Cavalerie*. The move from the M47 to the AMX30B was not perceived as positive to many experienced drivers. Driver training was quickly identified as a key element in the AMX30 crews' training cycle, and the aversion disappeared as newly trained conscripts without M47 experience arrived in the regiments. *(Fonds Claude Dubarry, Collection du Musée des Blindées de Saumur)*

RIGHT An AMX30B '430' (4th Squadron, 3rd Platoon, platoon commander's tank) was an early production tank serving in one of the RC70 regiments after 1984. The head-up position of the driver was standard operating procedure when manoeuvring in civilian areas, normally with the turret locked in the 12 o'clock position. *(Fonds Claude Dubarry, Collection du Musée des Blindées de Saumur)*

BELOW The basic equipment and layout of the French Army's AMX30B changed very little between the first production vehicles built in 1966 and the last built in 1981. The French armoured regiment changed from a five-tank platoon to the three-tank platoon at the same time that it adopted the AMX30B. These regiments, with four squadrons of 13 tanks (each of four platoons and a platoon command tank) and two tanks at regimental headquarters, were designated RC54 (with 54 tanks). This structure remained until 1984. *(Fonds Claude Dubarry, Collection du Musée des Blindées de Saumur)*

Unlike the Leopard, the AMX30B's armour was not upgraded during its service life and no improvements were made to its fire controls until the AMX30B2 was introduced in 1982. The decision to upgrade the tank was put off repeatedly in order to save money and to speed up the production of variants like the AMX30D, AMX30P and AMX30R. In comparison, the German Leopard saw some

The CN-105-F1 105mm gun had a maximum elevation of +20 degrees. The coaxial weapon (12.7mm machine gun and later the 20mm cannon CN-20-F2) could be decoupled and elevated separately to a maximum elevation of +40 degrees. *(Fonds Claude Dubarry, Collection du Musée des Blindées de Saumur)*

RIGHT The 501e RCC's '*Ney*', an early production AMX30B acting as OPFOR (opposing forces) during manoeuvres on the Mailly training ground. The turret is in the 6 o'clock position, possibly indicating that the tank is engaged in a delaying action, ready to move toward another defensive position. *(Fonds Claude Dubarry, Collection du Musée des Blindées de Saumur)*

incremental improvements in each production version that appeared after 1972 (such as main armament stabilisation), as did the American M60A1. The AMX30B was left largely in its original configuration, however, in part because the continued production of the tank in the 1970s was achieving a lower unit cost. This left the *Arme Blindée Cavalerie* of the late 1970s with a tank lacking the most modern features – features that GIAT had developed in the

meantime. The AMX30B's firepower, premised on the chemical energy of its hollow-charge anti-tank ammunition, was countered in some measure by laminate armour carried by Soviet tanks like the T64. The Obus-G as a doctrinal solution to destroying tanks, a basic principle behind the tactics developed for the AMX30B, was less of a 'silver bullet' by the mid-1970s. This did not change the army's orders. New-built AMX30Bs received by the army in the 1978–81 period still lacked gun stabilisation, a laser rangefinder and modern night vision equipment, differing only in their 20mm cannon secondary armament from the first batch built in 1966. This original configuration was retained for perhaps as many as 400 AMX30Bs still in service at the end of the Cold War. In the meantime, the army and GIAT proposed solutions while the government debated its military priorities.

BELOW The organisation of tank battalions between 1967 and 1984 included four tank squadrons and a mechanised squadron of infantry mounted in AMX10Ps. Mixing the infantry fighting vehicles with the tank squadrons for specific tactical scenarios was common. Here the *6e Régiment de Cuirassiers* were seen on the Mourmelon training area. The AMX10P in the foreground appears to be named '*Waterloo*' ... unusual for a French military vehicle! *(Fonds Claude Dubarry, Collection du Musée des Blindées de Saumur)*

RIGHT An early AMX30B during manoeuvres in a civilian area, which were frequent in the 1970s and 1980s. The 'camouflage paint' was a mixture of mud and diesel spread randomly on the tank by the crew. The NATO green paint scheme was retained by the French Army until 1984 and endured in many cases into the 1990s after the three-colour camouflage scheme was adopted. *(Fonds Claude Dubarry, Collection du Musée des Blindées de Saumur)*

AMX32: the tank the French Army could not afford

The AMX30B's fire controls represented the state of the art, and measured up well alongside any other fire control system in 1963. By the late 1960s improved electronics, ballistic computers and laser rangefinder technology had sufficiently advanced to be integrated into more accurate fire control systems.

The development of laser rangefinders and improved kinetic ammunition elsewhere also weighed against some of the older design features found in the AMX30B. The Yom Kippur War of 1973 made all the NATO nations re-evaluate tank design. If the anti-tank guided missile had made its mark, the experience of the Israeli Army in 1973 also placed a new emphasis on armour protection – and in particular, new forms of armour protection.

There were already plans to improve the AMX30's powertrain in the early 1970s. AMX30B GIAT documents dated 22 February 1973 spoke of the unsatisfactory life expectancy of the AMX30B's BV5 SD manual transmission, which had an average life of 2,500km before replacement was required. GIAT made great efforts to redesign or source an automatic

RIGHT The use of mud as camouflage was quite common among the tank crews before the adoption of a standard three-tone scheme. Note the blue box clipped on the main gun barrel – it is part of a SIMFIRE fire control simulator system. *(Fonds Claude Dubarry, Collection du Musée des Blindées de Saumur)*

ABOVE AMX30B from the second squadron of the *11e Régiment de Chasseurs* (11RCh) seen in 1980 during the Bastille Day parade in West Berlin. *(Fonds Claude Dubarry, Collection du Musée des Blindées de Saumur)*

transmission in the following years. The army was adamant that they required a transmission with a service life of 8,000km, and ideally 10,000km. They also plainly requested an automatic unit for greater ease of driving, availability within five years and suitability for retrofitting into the existing AMX30B design. As a result, several different transmissions were tested out, the best of which was the Minerva ENC 200.

By 1975 GIAT had identified a list of the AMX30B's other features that needed to be improved or replaced. The M208 coincidence rangefinder, for example, was very accurate to 2,000m, but optical fire controls only took account of distance to target. Parameters like the speed of the target, atmospheric temperature, wind direction and velocity and the temperature of the ammunition were essentially left to chance. The lack of a main armament stabiliser was a serious disadvantage that was never adequately rectified.

The OB-17-A and OB-23-A infrared sights could not be used in conjunction with the

ABOVE The NATO tactical sign on the left front mudguard of this AMX30B identifies it as part of the 2nd squadron, *4e Régiment de Cuirassiers* (4RC), *5e Division Blindée* (5th Armoured Division). The open ration boxes on the top of the turret suggest a break in training. The TOP7 cupola's excellent all-round vision and independent traverse and contrarotation offered a hunter-killer capability to a well-coordinated turret crew. *(Fonds Claude Dubarry, Collection du Musée des Blindées de Saumur)*

BELOW Large manoeuvres on and outside military training grounds were common in France and West Germany until the mid-1990s. The presence of this AMX30B in the middle of a sunflower field shows the comparatively small size and low profile of the tank, rendering it quite inconspicuous from the ground. Such a position could easily be spotted from the air, however! *(Fonds Claude Dubarry, Collection du Musée des Blindées de Saumur)*

ABOVE An overhead view of an AMX30B during manoeuvres. The half-round object above the camo net is the cover of the cupola ammunition tray. The ANF1 was loaded in this case with a 50-round 7.62mm blanks belt. The cupola tray could be loaded with up to 1,950 rounds. The two rings on the gun barrel probably indicate the platoon or squadron. *(Fonds Claude Dubarry, Collection du Musée des Blindées de Saumur)*

LEFT The combat snorkel was rarely seen fitted to the AMX30B during training. River crossing operations were dangerous and the training snorkel tower was usually employed on such occasions. These were large enough to provide crew members with a means of escape in case the engine cut out, but were too large to be carried on the tank. Due to the specialised training required for river crossing, only half of the AMX30B regiments were qualified in these operations, keeping in mind that the crews were normally conscripts. *(Fonds Claude Dubarry, Collection du Musée des Blindées de Saumur)*

LEFT An AMX30B stuck fast on a Cold War-era training area. Even with its light weight and good power-to-weight ratio, the AMX30B was as vulnerable as any other tank to ditching. *(Archives CAAPC)*

LEFT Getting ditched was often the result of the limited performance of the driver's infrared vision equipment, or simply inexperience.
(Archives CAAPC)

BELOW It was preferable for the tank commander to request help from another tank from the same platoon. Calling for the AMX30D of the support company was sure to result in reproachful comments ... and would always cost the crew a round of drinks!
(Archives CAAPC)

optical rangefinder and as a result the AMX30B had a limited effective combat range at night. Infrared sights were, by the later 1970s, far less effective than available image intensification equipment appearing on contemporaries like the M60A1 RISE Passive. The deployment of the feared T-64 and T-72 'super tanks' in the Soviet armoured forces in Eastern Europe added urgency to the need to replace the AMX30B.

The AMX32 was developed from 1975 as a vastly improved and more easily manufactured development of the AMX30 for only a moderate weight increase. It followed the AMX30's layout, using the same chassis and engine but with a new automatic transmission. The hull and

ABOVE By the late 1970s in France the modernisation of the AMX30B had been delayed repeatedly in the name of other military priorities. The AMX32 was first presented in 1979, but had been under development since the middle of the decade. Six AMX32 hulls and at least two turrets were constructed – the example seen here in 1979 was equipped with the 105mm gun in a welded turret. *(Nexter)*

RIGHT In France the laser rangefinder was first incorporated into the fire controls of the ACRA missile launcher included in the T142 turret for the AMX30 ACRA. The first practical application of a computerised fire control system was evaluated for inclusion in a tank turret by GIAT between 1975 and 1977. This system entered service as the COTAC (tank automatic fire control) in 1979 in the AMX10RC. The first tank equipped with the COTAC system was the AMX32, first shown at the Satory VII arms exhibition in 1979. *(Nexter)*

turret were welded, however, permitting spaced armour to be incorporated into the design. By using plates of differing hardness, improved protection was afforded against both high-explosive anti-tank chemical energy rounds (HEAT) and kinetic energy armour-piercing (KE) munitions on the frontal arc. In addition, it is believed that the armour was tested to provide immunity to 75mm *Perforant Coiffé à Ogive Traceur* (PCOT – an armour-piercing capped tungsten-cored tracer round developed for the AMX13's CN-75-50 gun rounds). The spaced armour was integral to the hull front, turret sides and gun mantlet (although the 105mm gunned turret seen on the first prototype lacked the spaced-armour mantlet).

A total of six AMX32 prototype hulls were built at ARE between 1979 and 1985. These

were tested extensively between 1979 and 1986. The decision to use the AMX32 to test improvements for the AMX30B upgrade was taken early on because documents from the beginning of 1980 speak of the existing order for AMX30B2s. Development continued in the hope of selling the AMX32 on the export market after the French Army rejected it on the grounds of unit cost and decided to adopt the cheaper AMX30B2 option. The AMX32's development offered a range of potential improvements to the AMX30 platform, which included an entirely new turret with advanced electronic fire controls, a

ABOVE The first AMX32 configuration shown was armed with the CN-105-F1, which could fire the range of munitions employed with the AMX30B as well as a new *Obus Flèche* APFSDS round. Ammunition capacity was 47 rounds stowed very similarly to the layout employed in the AMX30B. *(Collection Jerome Hadacek)*

vastly improved powertrain and possibly even a new main armament.

The AMX32 retained the basic suspension layout of the AMX30B but employed a reinforced torsion bar suspension to cope with a greater overall weight. It employed

LEFT The AMX32's spaced armour was a substantial improvement over the AMX30B's protection and was superior to other vehicles of similar weight. *(Nexter)*

RIGHT The second welded turret built for the AMX32 included a spaced-armour welded mantlet, mounting a CN-105-F1. This weapon was succeeded by the far more powerful CN-120-G1 120mm gun. Combined with the excellent new fire controls and the commander's panoramic sights, the AMX32 offered truly lethal firepower. *(Collection Jerome Hadacek)*

BELOW Bearing the *immatriculation* 994-0001, this AMX32 was extensively tested out by the STAT and the DGA. *(Collection Jerome Hadacek)*

AMX30B suspension components with five road wheel stations on each side, and the existing single-pin track. The AMX32 was offered with a modified Hispano-Suiza V12 HS-110-2 SR engine boosted to produce 720hp and subsequently 800hp. The gearbox first proposed was an AMX-designed 4AD gearbox which proved fragile, and this was then replaced by the BV5 SD manual gearbox employed on the AMX30B. Finally, the Minerva ENC 200 automatic transmission was fitted (the same type subsequently adopted for the AMX30B2). The later AMX32 prototypes were fitted with live-type double-pin track, which had been under development since 1973.

The AMX32 was the first French tank to mount the 120mm CN-120-G1 smoothbore gun, which was fitted to the second prototype turret in a special spaced-armour mantlet. The 120mm smoothbore was developed by

the *Etablissement d'Etudes et Fabrications d'Armement de Bourges* (EFAB), the renamed Bourges arsenal which had perfected the CN-105-F1. The chrome-lined, auto-frettaged CN-120-G1 measured 7.15m, and weighed 2,620kg (including its elevation gear). It was designed to fit the same trunnion width as on the existing AMX30B turret, and it was theoretically capable of being mounted in an upgraded AMX30B (where it was test fired prior to the manufacture of the first AMX32 turret). Maximum chamber pressure was quoted at 6,300 bars. It was designed for compatibility with the range of ammunition designed for the Rheinmetall L44 120mm smoothbore gun. The CN-120-G1 fired an *Obus Flèche* (which translates from the French as 'arrow round') with 5.8kg tungsten penetrator at 1,700m/sec to a range of 2,000 to 3,000m. The 120mm *Obus Charge Creuse* (or HEAT) projectile weighed 13kg and was fired at a muzzle velocity of 1,100m/sec. These rounds could easily penetrate NATO heavy tank targets at ranges of 3,500m. Both complete rounds weighed around 24kg. The CN-120-G1 had a relatively long recoil stroke of 485mm which minimised its recoil force to 36,000kg, which made it suitable for fitting on a tank in the 37–40-tonne range.

Like the CN-105-F1, the 120mm gun was fitted with a magnesium alloy thermal sleeve and employed a compressed air purge system for fume extraction rather than the barrel-mounted extractor employed on the West German gun. The 120mm ammunition was stowed in the hull front next to the driver and in the turret (with a total of 38 rounds stowed). The CN-120-G1 could also theoretically fire all ammunition types employed by the Rheinmetall 120mm L44 gun. All of the AMX32 prototypes were fitted with the CN-20-F2 secondary armament.

The AMX32 was the first French tank design capable of electronic 'hunter-killer' target acquisition, a capability in 1979 only available on two new Western battle tank designs – the American M1 and the West German Leopard 2. The AMX32's gunner aimed through his M581 telescopic sight installed in the right-hand side of the gun mantlet. The tank commander's cupola (an improved and more diminutive version of the TOP7) mounted an independently traversable M527 panoramic sight gyroscopically stabilised to the line of sight. The main armament could be slaved to the M527 site for firing on the move, but the gun itself was not stabilised. The M581 sight incorporated the APX CILAS M550 laser rangefinder and had 10× magnification. The M527 and M581 sights were integrated into the COTAC (*COnduite de Tir Automatisée pour Char*) automatic fire control system. This system employed an APX M241 ballistic computer, which positioned the gunner's sight reticule and determined all

BELOW An AMX32 prototype under test on 6 November 1984. By this time the AMX30B2 had been adopted and was in production at ARE. *(Collection Jerome Hadacek)*

ABOVE An AMX32 with its full complement of ammunition and onboard stowage. This vehicle is armed with the 120mm CN-120-G1. *(Nexter)*

BELOW Seen on a manoeuvre area, AMX32 994-0001 is under test in 1985. The French Army could not afford this excellent tank, and no foreign orders were forthcoming. *(Thomas Seignon)*

necessary range, trunnion position and attitude data. Night vision was provided by means of the Thomson-CSF DIVT13 low-light television camera, an image-intensification system far superior to the OB-17-A and OB-23-A infrared systems standardised on the AMX30B.

The AMX32 was seriously considered by the French Army as a potential replacement for the AMX30B during its conceptual stages, but this notion was dropped in 1980 due to budgetary constraints. The AMX32 was used instead to prove the concepts that would be embodied in the much cheaper AMX30B2, excepting, of course, the latter's 120mm gun. Nonetheless the AMX32 was considered to be an ideal MBT type to sell on the export market, particularly in the Middle East. GIAT's successes in selling the AMX30S and derivatives in Saudi Arabia, Qatar and in the United Arab Emirates (and in selling the AU F1 in Iraq) led to high hopes for selling the AMX32 in turn. None of these sales came to pass, however, and in 1987 GIAT's efforts to

market the AMX32 ceased as the even more advanced AMX40 was designed – though it too was destined never to sell.

Renewal: the AMX30B2 development, production and conversion programmes, 1980–91

The first and only true upgrade applied to the AMX30B production line was the installation of the new CN-20-F2 20mm coaxial weapon in 1974. This was followed by a base overhaul programme to integrate the 20mm weapon into vehicles built prior to 1975. After this, however, further development was weighed against the purchase of a new battle tank. This was an agonising process which was delayed by government bureaucracy, changes in equipment priority and budgetary shortfalls. A new tank was beyond the army's budget at the end of the 1970s, and the AMX30B was still a relatively new vehicle. When the army opted to upgrade the AMX30B comprehensively instead of buying the AMX32, they decided that the most effective improvements to introduce would be to adopt the turbocharged HS-110-2 engine, the ENC 200 automatic transmission and upgrade the turret with the most cost-effective elements of the AMX32's COTAC fire control systems. This upgraded tank was baptised AMX30B2, and production was undertaken in 1981.

Alongside the last batch of AMX30Bs built for French orders, a first order of nine AMX30B2s (*Tranche 1*, which included the test hull, five *Chars de Rang* and three *Chars Peloton*) were assembled at ARE – converted from available AMX30B hulls and turrets diverted from the main assembly line. The first complete tank was presented to the army in November 1981, a

ABOVE The definitive AMX30B2 prototype under evaluation with a STAT crew. Numbered 654-0097, it mounted a modified T105 turret with its gun mantlet prominently signed for 'Danger: Laser'. Five such vehicles were manufactured to test out the AMX30B2 concept in a platoon-sized unit. Some details were not yet finalised, such as the armoured cupola mounting for the 7.62mm ANF1 machine gun. The AMX32's cupola and commander's panoramic sights were not adopted for the AMX30B2, but the COTAC system fit readily into the T105 turret. *(STAT)*

second in December. The remaining six were all completed in early 1982. These vehicles were successfully evaluated, two by the STAT, another by the gunnery establishment at Bourges, and five delivered directly to a special trials platoon formed by the 503e RCC at Mourmelon. The AMX30B2 represented a very good balance of economy and improvement. The incorporation of modern fire controls, the improved suspension system and the automatic transmission resulted in a much improved AMX30 design.

OPPOSITE The CN-105-F1 gun was unchanged in the AMX30B2 programme, thanks to DEFA's exacting gun specification of 1964. The gun was still in production when the AMX30B2 was ordered into production. On the tanks converted from AMX30Bs each gun was stripped down and rebuilt. *(Jerome Hadacek)*

RIGHT The AMX30B2 definition prototype numbered 654-0097 was revealed to the public in 1983. The original armour was unchanged, in part to improve the type's mobility without additional weight, but also because the AMX30's extensive use of cast armour on its front hull and turret did not lend itself particularly well to additional armour. *(Jerome Hadacek)*

RIGHT The AMX30B2 prototype seen from the rear. Two differences from the AMX30B become immediately clear: the enlarged turret bustle box (which now held the updated NBC system) and the central position of the AT17 telephone box on the hull rear plate, between the two jerrycan mountings. *(STAT)*

A second batch, comprising 54 vehicles, was produced in 1983, with a third of 60 in 1984–85 and a fourth of 44 in 1986–87. In all 166 newly built AMX30B2s were delivered to French orders. These vehicles were all fitted with the DIVT13 image-intensification camera in a sealed watertight barbette on the right-hand side of the mantlet. The DIVT13 was a substantial advance on the older infrared kit that had equipped the AMX30B, allowing the crew to identify and engage targets at longer range than had previously been possible. The old infrared driving lights and mantlet projector were retained on the AMX30B2 as a secondary system and as a white light searchlight respectively, continuing in service right into the 1990s.

RIGHT From the front the AMX30B2 could be distinguished from the AMX30B by the absence of the optical rangefinder 'ears' on the turret sides and the mounting for the DIVT13 low-light television camera on the right-hand side of the mantlet. *(Fonds Claude Dubarry, Collection du Musée des Blindées de Saumur)*

LEFT This a squadron commander's tank, one of the first batch of newly built AMX30B2s issued to the *503e Régiment de Chars de Combat* (RCC) in 1985 or 1986. The 503e RCC were completely equipped with AMX30B2s between 1982 and 1985. At the same time, the regiment established the tactical viability of the four-tank platoon, introduced as part of the reorganisation of 1984. *(Hugues Acker)*

For the later AMX30B2 production batches, economy was stressed as far as possible. Existing AMX30Bs were sent back to ARE in batches for rebuilding, a process which required a complete disassembly of the vehicle amounting to a full factory rebuild. The hull of each tank was gutted, its engine and transmission removed and replaced with a rebuilt HS-110-2 turbocharged engine, its associated cooling system and the Minerva EN-200 automatic transmission. The rear hull plate was replaced. The driver's position was updated with a new steering wheel in place of the AMX30B's tiller bars. The suspension was also removed and replaced, and original components were recovered to be rebuilt and reused wherever possible. The turret was treated similarly, with the armament removed and rebuilt and the original fire controls, NBC and radio systems taken out. The turrets were then completely re-equipped with new sights and electronic fire controls. The radio layout, which extended throughout the turret in the AMX30B, was grouped behind the commander's position where the NBC system had been located. The NBC system itself was removed and relocated to a large box attached to the exterior of the turret bustle. Records indicate that turrets, 105mm and 20mm guns, engines and hulls normally did not match to their original pairings during remanufacture.

The adoption of high-velocity kinetic ammunition for the CN-105-F1 proved to be an

ABOVE This AMX30B2 was the reference tank retained by the STAT. The DIVT13 is shown with the camera shutter plate in the open position, as it would appear in use. *(STAT)*

LEFT This AMX30B2 was one of the 166 newly built tanks, identified by the large size of the gunner's sight shutter. The PH8B searchlight was retained on the left side of the mantlet. The PH8B was retained in service as a white light searchlight into the 1990s. *(STAT)*

RIGHT This AMX30B2 (lacking a visible registration number) carries a three-tone camouflage paint scheme (although not quite to the official 1984 standard, which was supposedly identical for each type of army vehicle). The right-side aperture shield on the camera box indicates that this is a late production vehicle, fitted with a DIVT16 thermal camera. Despite its new electronics, the AMX30B2 was fully sealed for submerged river crossing just like its predecessor. *(STAT)*

important change in French anti-tank gunnery doctrine. While it was never widely publicised by DEFA or DTAT (especially in the early 1960s when the Obus-G was being perfected and became practically enshrined in French military dogma), KE rounds were considered as a viable alternative for the D1512/CN-105-Mle 62/CN-105-F1 from the time of its conception. Efforts to develop a functional kinetic energy armour-piercing round with a sub-calibre penetrator were championed within DEFA by *Ingenieur Général de l'Armement* Maurice Carrougeau. The most important development of this type was the 105/60/43mm projectile, set aside when Carrougeau retired. The research conducted was useful in the development of armour-piercing fin-stabilised discarding sabot ammunition in 105mm calibre when KE armour-piercing rounds once again found favour for the improvement of the AMX30's firepower.

The AMX32 had already been used to prove this ammunition with the CN-105-F1 and the COTAC system. The OFL-F1 (*Obus Flèche F1*) round was fired at 1,525m/sec. It could be stowed in the AMX30B2 alongside the existing ammunition types developed for the

ABOVE The DIVT16 CASTOR thermal television camera was developed by Thomson-CSF (and later Thales) for the AMX40 in the 1985–87 period. It was adopted for the last batches of AMX30B2s converted from existing AMX30Bs, fitting neatly into the place previously occupied by the DIVT13 low-light television camera on the right side of the gun mantlet. It was succeeded by the simpler DIVT18 system on about half of the AMX30B2 Brennus tanks, although the mounting was identical to that seen here. The yellow decal on the main gunnery sight cover in the mantlet is a laser warning. *(Pierre Delattre)*

RIGHT This AMX30B2 seen at the STAT in the late 1980s shows standard stowage practice for the type. Later converted vehicles were distinguished by the rangefinder ports welded over with circular plates. The AMX30B2 was extremely well regarded in the *Arme Blindée Cavalerie*, notwithstanding its lack of a main armament stabiliser, and despite its limited firepower in comparison to more modern designs armed with 120mm guns. *(STAT)*

CN-105-F1. The optimum range when firing the OFL-F1 was of course much shorter than with the OCC-Mle 61 Obus-G round, with an effective range of 1,600m. This was typical for other KE armour-piercing rounds of the time and was limited by the rifling retained for the CN-105-F1. The penetrator adopted for the OFL-F1 was adapted from the type developed for a high-velocity 90mm round introduced in the late 1970s.

The AMX30B2's chassis was strengthened in light of the experience gained with the development of the AMX32 hull. The torsion bars employed with the upgraded suspension were increased in diameter from 53mm to 55mm, and each bar was wrapped in rubber rather than the original neoprene. The swing arms fitted to the first and fifth road wheel stations on each side were of a new type and were not interchangeable with the AMX30B. The double-action Messier oscillating dampers adopted for the AMX30B2 suspension were specified to provide 1.4× more damping torque than the original pattern. Combined with the automatic transmission, the new suspension gave the AMX30B2 a smoother ride and rendered it a very stable gun platform. The AMX30-type track was retained for the AMX30B2 but after 1990 a new double-pin track system (first tested in 1973) was supplied. The new tracks required a different sprocket ring.

Four *tranches* of AMX30B2s were converted

LEFT The LIR 30 was intended to interrupt the infrared beam used to control anti-tank guided missiles, and is seen in operation here. *(STAT)*

LEFT The new Galix smoke grenade launchers were adopted at the same time (late 1990) for the AMX30B2s being prepared for service in the Gulf with the *4e Régiment de Dragons*. Range periods firing live OFL-F1 (APFSDS) ammunition were also part of the STAT evaluations prior to deployment. *(STAT)*

by the main assembly line at GIAT for a total of 236 tanks. These included the *5e Tranche* (73 vehicles) ordered in the 1984 budget and delivered in early 1987, the *6e Tranche* (50 vehicles) delivered in 1988–89, the *7e Tranche* (50 vehicles) delivered in 1989–90 and the *8e Tranche* (63 vehicles) starting in 1990. In 1991 a final order for 93 AMX30B2 conversions from ARE was approved in the national assembly and these were later cancelled. A secondary AMX30B2 conversion line was set up at the army workshops at Gien, but the number of conversions it completed has never been publicly disclosed. The cost of an AMX30B2 conversion was estimated at 7 million Francs, including its turret electronics, its rebuilt engine and its new automatic transmission – a much lower price than what it would have taken to buy an AMX32. It is believed that over 600 AMX30Bs were converted into AMX30B2s

BELOW The provision of two 200-litre fuel barrel brackets that could be ejected from the driver's position (later standard on the Leclerc) was tested for the AMX30B2 by the STAT. This kind of solution was nothing revolutionary, having been previously tested for the M47 in the 1960s! *(STAT)*

LEFT In the 1990s the French Army was reduced considerably. The AMX30B2 was retained in first-line service while the Leclerc was field tested and was used to prepare units for conversion to the new tank. Stocks of 105mm and 20mm ammunition were abundant and the conversion process was normally preceded by extensive gunnery and tactical training. Thereafter, the best remaining AMX30B2s were gathered for use by the FORAD, like this example captured for posterity during Operation Azure, an urban exercise in north-eastern France in 2006. *(Pierre Delattre)*

BELOW The FORAD AMX30B2s were for many years operated by the *5e Régiment de Dragons*, in two OPFOR groups (at Mailly and at Sissonne). By 2006 many of these tanks were repainted in grey with black disruptive stripes, with sheet metal added to better disguise them as 'Soviet-built' vehicles. Attrition set in within seven or eight years, and by 2014 most of the HS-110-2-engined vehicles had run short on parts and were retired. These grey and black FORAD tanks were from the detachment based at Mailly. *(Jerome Hadacek)*

between the ARE and Gien production lines. The remaining AMX30Bs remained in service with the French Army alongside the AMX30B2 until 1997.

AMX30B2 Brennus: the last upgrade

The AMX30B2 conversion programme really hit its stride just as the Cold War ended, at which time there was every intention of eventually converting all of the old AMX30Bs. After the cessation of the Cold War in 1990, it became painfully clear that the army had to be reduced, and since the new AMX56 Leclerc was not yet in service, it was decided to upgrade the AMX30B2's powertrain once more. The American Mack E9 V8 diesel engine was chosen to replace the HS-110 in most of the AMX30 family types and was expected to be retained in service after 1995. These included the AMX30B2, the AU F1 and the EBG – but not the AMX30D or the AMX30 Roland, which required specific power take-off equipment to

ABOVE The first Brennus conversion was a proof of concept undertaken by GIAT on an original 1984 production AMX30B2 equipped with the DIVT13 image intensification low-light television camera and original T105M cast turret. The layout of the reactive armour certainly cluttered the already diminutive AMX30B2 and caused a certain amount of the glacis and turret exterior stowage to be relocated. *(STAT)*

BELOW Notably, the AMX30B2 Brennus demonstrator had the battery covers relocated to the engine deck, retaining its deep fording capability. The demonstrator was rebuilt in 1996 with a DIVT18 camera and was issued in the final conversion batch. *(STAT)*

ABOVE A series of spaced mounting stubs were welded to the armour. The mounting frames were attached to these. On the turret this required precise measurement to prevent the ERA bricks from seating too close to the armour. *(STAT)*

BELOW Here we can see the mounting frames fixed on each side of the turret, on the glacis, on the gun mantlet and on the front slope of the turret roof. *(STAT)*

operate winches and electronics. Manufactured in the USA, the Mack engine was chosen 'off the shelf' because it fitted AMX30's existing engine compartment and was compatible with the ENC200 automatic transmission. Some 500 E9 engines were delivered to Limoges for modification with a dry sump system in 1994, and then on to Gien where the installations were undertaken. The first Mack-engined AMX30B2s were issued to the *5e Régiment de Dragons*.

The experience of Operation Daguet (a Daguet is a type of antelope or gazelle, used as the French Army's codename for their role in the larger-coalition Operation Desert Storm) and the possibility of deploying armoured units on peacekeeping duties into the Balkans in the 1990s forced the army to reconsider the AMX30B2's armour protection. GIAT had developed its own explosive reactive armour (ERA) system during the 1980s, but the French Army did not place any orders until after the First Gulf War. Known as the *Briques de Surblindage* G2 (BS G2), these conferred

BELOW LEFT The BS G2 system consisted of a hardened steel plate mounted to the frame (seen on the bottom row at the lower edge of the turret side). The plates each carried an ERA brick, comprised of two steel plates sandwiching an explosive charge. *(STAT)*

BELOW The additional armour required the driver's periscopes to be raised and a hinged ERA frame section on the lower left of the gun mantlet was fitted with a pivoting hinge (enabling it to be swung outward to permit easy passage for the driver to his hatch). The battery compartment covers were relocated to the engine decks. *(STAT)*

FAR LEFT The inner plate, with its cast body. *(STAT)*

LEFT The inert version of the GIAT BS G2 was designated BS G1, and was in fact the type carried by all AMX30B2 Brennus tanks in peacetime. This mounted on top of the inner plate by means of threaded mounting bolts. *(STAT)*

the same level of protection as 400mm-thick Rolled Homogeneous Armour (RHA) plate at a 60-degree incidence – for a weight penalty of roughly 1 tonne. The modified vehicles were named AMX30B2 Brennus, giving the army an MBT suitable for deployment into a peacekeeping role in emergency. In peacetime the tanks carried BS G1 inert ERA bricks.

A single Brennus 'prototype' (demonstrator might be a better choice of words) was prepared after Operation Daguet from an early production AMX30B2. After trials to prove that the additional armour did not affect mobility or deep fording capability, an order for 80 conversions proceeded. The first AMX30B2 Brennus conversions were undertaken in two blocks during 1995 on AMX30B2s from the late conversion batches. All were fitted with the DIVT16 CASTOR thermal gunnery camera. The 1995 conversion batches included 26 HS-110-powered vehicles and a further 25 vehicles that had been fitted with the Mack E9 engine.

Most of the HS-110-powered Brennus conversions were tanks that had been turned in for rebuilding by the *4e Régiment de Dragons* (after combat service in Operation Daguet). The E9-engined conversions had been turned in by the *5e Régiment de Dragons*. The AMX30B2 Brennus conversions included old AMX30Bs rebuilt into AMX30B2s between 1987 and 1990 and a few of the last AMX30B2s newly converted in 1990. These vehicles carried the most advanced technological features incorporated into any AMX30B2 conversion batches. They featured the LIR30 infrared lure system, the Galix smoke grenade-launching

system and the DIVT16 thermal camera system. The first AMX30B2 Brennus conversions were issued to the *4e Régiment de Dragons* and the *2e Régiment de Chasseurs* to equip full tank squadrons. Once equipped with GIAT BS G1 reactive bricks, only their manufacturers' plates and *immatriculation* numbers indicated their precise vintages.

The second block of 20 Mack E9-powered AMX30B2s and 10 HS-110-engined AMX30B2s with hydrostatic cooling (including the reworked prototype 6844-0056) were converted in 1996.

BELOW An AMX30B2 Brennus under evaluation by the STAT. The conversions were issued to the professionalised squadrons in the *4e Régiment de Dragons* and *501e Régiment de Chars de Combat* but the Leclerc's arrival saw them withdrawn very quickly into a single unit. The last cavalry regiment equipped with the AMX30B2 in the French Army was the *1e–2e Régiment de Chasseurs* (1e–2e RCh, based at Verdun), which operated all of the 81 Brennus conversions by 1998. The type was withdrawn from service in 2006 and some 40 of the E9-engined vehicles were placed in storage for eventual use by the FORAD. *(STAT)*

1995 Batch				
Serial	Conversion date	Year of original manufacture	Engine type	Thermal camera type
6380285	1995	1967	Renault Mack E9	DIVT16 CASTOR
6380293	1995	1967	Renault Mack E9	DIVT16 CASTOR
6780223	1995	1968	Renault Mack E9	DIVT16 CASTOR
2940084	1995	1969	Renault Mack E9	DIVT16 CASTOR
2940123	1995	1969	Renault Mack E9	DIVT16 CASTOR
2940135	1995	1969	Renault Mack E9	DIVT16 CASTOR
6480012	1995	1969	Renault Mack E9	DIVT16 CASTOR
6040346	1995	1970	Renault Mack E9	DIVT16 CASTOR
6140089	1995	1971	Renault Mack E9	DIVT16 CASTOR
6140191	1995	1971	Renault Mack E9	DIVT16 CASTOR
6140196	1995	1971	Renault Mack E9	DIVT16 CASTOR
6240105	1995	1972	Renault Mack E9	DIVT16 CASTOR
6240108	1995	1972	Renault Mack E9	DIVT16 CASTOR
6240135	1995	1972	Renault Mack E9	DIVT16 CASTOR
6340102	1995	1973	Renault Mack E9	DIVT16 CASTOR
6340129	1995	1973	Renault Mack E9	DIVT16 CASTOR
6340135	1995	1973	Renault Mack E9	DIVT16 CASTOR
6340174	1995	1973	Renault Mack E9	DIVT16 CASTOR
6640142	1995	1976	Renault Mack E9	DIVT16 CASTOR
6640164	1995	1976	Renault Mack E9	DIVT16 CASTOR
6640168	1995	1976	Renault Mack E9	DIVT16 CASTOR
69040103*	1995	1990	Renault Mack E9	DIVT16 CASTOR
69040106*	1995	1990	Renault Mack E9	DIVT16 CASTOR
69040107*	1995	1990	Renault Mack E9	DIVT16 CASTOR
69040112*	1995	1990	Renault Mack E9	DIVT16 CASTOR
6480007	1995	1967	HS-110 Hydrostatic	DIVT16 CASTOR
6480010	1995	1967	HS-110 Hydrostatic	DIVT16 CASTOR
6780213	1995	1967	HS-110 Hydrostatic	DIVT16 CASTOR
2840182	1995	1968	HS-110 Hydrostatic	DIVT16 CASTOR
2840189	1995	1968	HS-110 Hydrostatic	DIVT16 CASTOR
6880209	1995	1968	HS-110 Hydrostatic	DIVT16 CASTOR
2940080	1995	1969	HS-110 Hydrostatic	DIVT16 CASTOR
6040230	1995	1970	HS-110 Hydrostatic	DIVT16 CASTOR
6040253	1995	1970	HS-110 Hydrostatic	DIVT16 CASTOR
6040334	1995	1970	HS-110 Hydrostatic	DIVT16 CASTOR
6040338	1995	1970	HS-110 Hydrostatic	DIVT16 CASTOR
6140095	1995	1971	HS-110 Hydrostatic	DIVT16 CASTOR
6140195	1995	1971	HS-110 Hydrostatic	DIVT16 CASTOR
6140200	1995	1971	HS-110 Hydrostatic	DIVT16 CASTOR
6240179	1995	1972	HS-110 Hydrostatic	DIVT16 CASTOR
6340093	1995	1973	HS-110 Hydrostatic	DIVT16 CASTOR
6340109	1995	1973	HS-110 Hydrostatic	DIVT16 CASTOR
6340124	1995	1973	HS-110 Hydrostatic	DIVT16 CASTOR
6340152	1995	1973	HS-110 Hydrostatic	DIVT16 CASTOR
6440078	1995	1974	HS-110 Hydrostatic	DIVT16 CASTOR
6640145	1995	1976	HS-110 Hydrostatic	DIVT16 CASTOR
6640162	1995	1976	HS-110 Hydrostatic	DIVT16 CASTOR
6640166	1995	1976	HS-110 Hydrostatic	DIVT16 CASTOR
6640188	1995	1976	HS-110 Hydrostatic	DIVT16 CASTOR
6640190	1995	1976	HS-110 Hydrostatic	DIVT16 CASTOR
6640191	1995	1976	HS-110 Hydrostatic	DIVT16 CASTOR

1996 Batch				
Serial	Conversion date	Year of original manufacture	Engine type	Thermal camera type
6480024	1996	1967	Renault Mack E9	DIVT18
2840173	1996	1968	Renault Mack E9	DIVT18
2940088	1996	1969	Renault Mack E9	DIVT18
6140092	1996	1971	Renault Mack E9	DIVT18
6140171	1996	1971	Renault Mack E9	DIVT18
6140183	1996	1971	Renault Mack E9	DIVT18
6240109	1996	1972	Renault Mack E9	DIVT18
6240110	1996	1972	Renault Mack E9	DIVT18
6340105	1996	1973	Renault Mack E9	DIVT18
6340115	1996	1973	Renault Mack E9	DIVT18
6340122	1996	1973	Renault Mack E9	DIVT18
6540097	1996	1975	Renault Mack E9	DIVT18
6640125	1996	1976	Renault Mack E9	DIVT18
6640165	1996	1976	Renault Mack E9	DIVT18
6640248	1996	1976	Renault Mack E9	DIVT18
69040068*	1996	1990	Renault Mack E9	DIVT18
69040104*	1996	1990	Renault Mack E9	DIVT18
69040110*	1996	1990	Renault Mack E9	DIVT18
69040113*	1996	1990	Renault Mack E9	DIVT18
69040114*	1996	1990	Renault Mack E9	DIVT18
6280010	1996	1967	HS-110 Hydrostatic	DIVT18
2840130	1996	1969	HS-110 Hydrostatic	DIVT18
6480289	1996	1969	HS-110 Hydrostatic	DIVT18
6140090	1996	1971	HS-110 Hydrostatic	DIVT18
6640143	1996	1976	HS-110 Hydrostatic	DIVT18
68440056**	1996	1984	HS-110 Hydrostatic	DIVT18
69040105*	1996	1990	HS-110 Hydrostatic	DIVT18
69040108*	1996	1990	HS-110 Hydrostatic	DIVT18
69040109*	1996	1990	HS-110 Hydrostatic	DIVT18
69040111*	1996	1990	HS-110 Hydrostatic	DIVT18

* vehicles built as AMX30B2
** prototype test of concept

RIGHT The BS G2 armour system had very little impact on the AMX30B2 Brennus's mobility due to its extremely light weight. On this vehicle we can see the hinged section of ERA bricks over the driver's position swung open to permit him to embark. (Pierre Delattre)

These vehicles differed from the 1995 batches in that they carried the simpler DIVT18 thermal camera (which could not be distinguished from the earlier DIVT16 CASTOR from the exterior). In all, 81 AMX30B2 Brennus conversions were issued, the last of which entered service during 1997. It should be pointed out that live BS G2 *quebriques* would only have been fitted in the event of a combat deployment, which never came. By the time of the 1998 Kosovo crisis the Leclerc was in service and the AMX30B2 Brennus enjoyed its ten-year active duty career without a single deployment.

RIGHT Of course, by the time of writing in 2019, spares and a lack of personnel trained in the repair of such old equipment has put their continued use into a finite number of hours. *(Thomas Seignon)*

OPPOSITE One of the series conversions under test by the STAT, equipped with the DIVT16 CASTOR thermal camera, the LIR30 system and Galix smoke grenade launchers. *(STAT)*

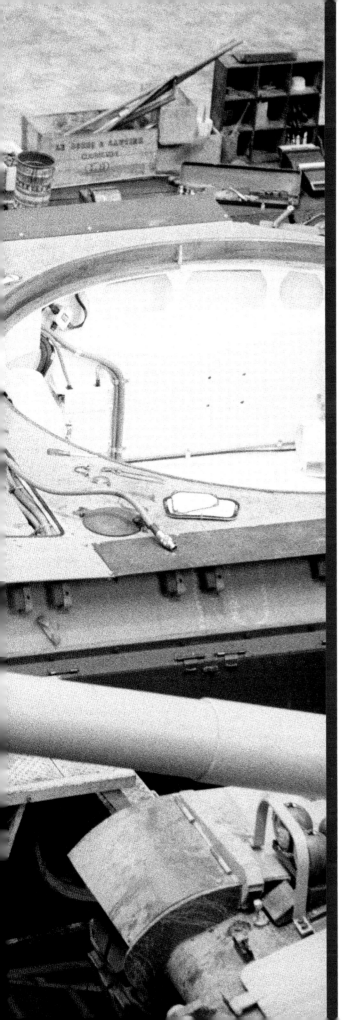

Chapter Two

Anatomy of the AMX30B

The AMX30B design provided its crew with a powerful main armament and modern fire controls for combat by day or night. It was sealed for submerged river crossings and for action on a contaminated battlefield. It was lightly armoured but mobile as a result. Despite its sophisticated features, the AMX30B design strove for simplicity and every subsystem was designed for a conscript crew.

OPPOSITE The AMX30 assembly line at ARE in the late 1970s. The AMX30 hull seen under construction is fitted with sand shields extending from the hull sides to minimise dust entering the engine air intakes (and is probably an AMX30S destined for export to Saudi Arabia). Exports differed from the standard AMX30B in a minimal number of subsystems (gun sights, and powertrain ratings optimised for use in hot, dusty climates). *(Getty Images, Gilbert Uzan)*

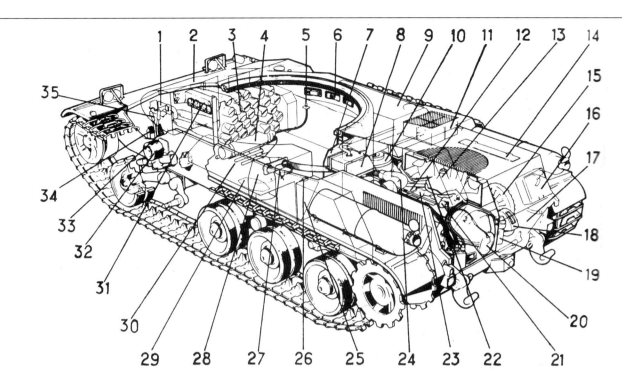

1 Driver's gear selector	**9** Right rear fuel tank	**16** Infantry telephone	**25** Left-side rear fuel tank
2 Front hull fuel tank	**10** Cooling system pump	**17** Lower hull transmission	**26** Turret relief valve
3 Front hull ammunition	**11** Right-side battery	inspection plate (hinged)	**27** Rotary junction, turret
stowage (28 rounds)	compartment, battery and	**18** Right-hand steering brake	electrical power
4 CO_2 bottle for fire	grille (left-side battery not	assembly (left hand not	**28** Air filter
suppression system	visible)	visible)	**29** Crew escape hatch
5 Gunner's fire suppression	**12** Hispano-Suiza HS-110	**19** Transmission oil filter	**30** Suspension
system	diesel engine	**20** Transmission	**31** Driver's dashboard
6 Control panel, fire	**13** Main cooling fan	**21** Odometer wheel	**32** Track return roller
suppression system	**14** Right-side radiator (left side	**22** Sulzer cooling system	**33** Driver's accelerator
7 Crew compartment/engine	not visible)	**23** Gearbox	pedal
compartment fire wall	**15** Right-hand final drive and	**24** Left-side turbo-compressor	**34** Driver's brake pedal
8 Left-side oil tank (right-side	main brake assembly	intake	**35** Driver's steering levers
tank not visible)			

ABOVE The AMX30B's hull as described in the first training manual (which used the two *chars de définition* of 1965 as the model in many instances). The hull underwent very few changes throughout AMX30B production. *(Collection Thomas Seignon)*

The production AMX30B was a relatively small and light battle tank. It measured 9.5m from the gun muzzle to the rear mudguard edges. The hull itself was 6.58m long and 3.05m wide. It presented a low silhouette, measuring 2.29m from the ground to the highest point of the turret roof. The TOP7 vision cupola added an additional 56cm to the tank's overall height but brought with it perhaps the best panoramic vision cupola of its time.

The AMX30B hulls were serialled in the 5000 and 6000 series, from the first hull, numbered 5001. Production hulls were welded together from flat plate and from castings with a maximum frontal armour thickness of 70mm. Hull manufacture was undertaken at ARE. The

driver's position and the glacis plate itself were castings welded into the larger assembly. A special turret traverse hydraulic stop switch was provided for the driver's protection when driving hatch-open.

The AMX30B's turret was designated T105 (for *tourelle 105mm*) and production turrets were numbered sequentially from 1 upwards. The original design was drawn up at the *Manufacture d'Armes de Saint-Étienne* (MAS), an establishment with a long history of designing weapons. Overall responsibility for the turret design was transferred to the *Atelier de Tarbes*, where they were subsequently manufactured. The prototype turrets developed from an oval-shaped turret somewhat

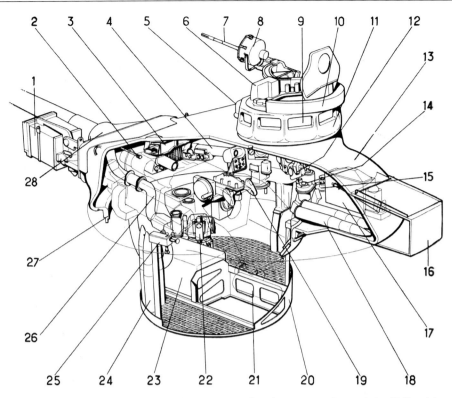

1 PH8A infrared/white light projector	7 ANF1 7.62mm cupola machine gun	14 Cupola contra-rotation control
2 Fume evacuation port	8 PH9A cupola infrared searchlight	15 Turret traverse lock
3 M208 coincidence rangefinder	9 M268 episcope	16 Turret stowage box
4 M271 gunner's sight	10 ANF1 7.62mm feed tray	17 NBC system
5 M282 (day)/OB-17-A (night) commander's sight	11 TOP7 cupola	18 Turret bustle ammunition stowage (18 rounds of 105mm)
6 Commander's binocular sight M267 (day)/OB-23-A (night)	12 Tank commander's position	19 Gunner's position
	13 *Corps pivotant* or pivoting portion of the M271 main gunner's sight	20 Turret basket
		21 Compressed air bottle (for breech fume evacuation system)

22 Turret traverse hydraulic motor	
23 TR-VP-13-A radio set installation	
24 Loader's ready rack (three rounds)	
25 Compressed air control and taps	
26 Fume evacuation pipe	
27 Ball-bearing race	
28 Gun mantlet	

reminiscent of a T-54's turret in 1960, to an elongated form on the AMX30As. It was further modified into its final form on the two *chars de définition* built in 1965. The SAMM S470 cupola was employed on all of the AMX30As and on one of the *chars de définition* because the TOP7 cupola went through a long development process at MAS and APX. Prototype and production turrets were all single-piece castings, with the thickest armour (80mm) on the front face. The mantlet pivoted on widely spaced trunnions, also with a maximum armour thickness of 80mm. It was pierced on the right-hand side for the gunner's telescopic sight and on the left for the coaxial armament, which was fitted into a detachable section designed for

ABOVE A simplified diagram showing the T105 turret layout derived from the *char de définition* **254-0745. The PH8A searchlight was dropped for this vehicle in favour of the PH8B, a lighter mounting, which equipped the AMX30B and the AMX30B2. Nonetheless this was the standard turret cutaway drawing for the early AMX30B manual. Note that the second radio set is omitted, among other details.** *(Collection Thomas Seignon)*

independent elevation. The shape of the mantlet casting was excellent from a ballistic point of view and incorporated seals at every joint and opening to permit waterproofing.

The turret was designed with good ergonomics for a high rate of fire. Traverse was hydraulic and was controlled by the gunner's hand control. A manual backup was also provided. Ammunition stowage was carefully distributed to permit the loader's access to as many rounds as possible. A three-round ready rack was located to the front left of the loader's seat. The turret bustle incorporated a larger ready rack with 16 rounds. The third ammunition rack was located on the right side of the driver behind the glacis (in front of the gunner). This held 27 rounds, which were accessed with the turret traversed to the right. The loader's radio installation was located immediately in front of his position on the turret basket floor plate.

Automotive drive train – engine/transmission/final drives, tracks and suspension

The original vision for the 30-tonne design, as set out in FINABEL 3A5 in 1957, was premised on the D1500 series 105mm gun design and the SOFAM 12GS DS 750hp petrol engine. The 30-tonne tank project was executed through DTAT's network of arsenals. The apparent exception to this rule was the license manufacture of the Hispano-Suiza HS-110 engine by SAVIEM (a division

of Renault – already by then a nationalised corporation under direct government control). Production was undertaken at Limoges in a factory operated in the former DEFA facility, also overseen by the state.

AMX, the design parent for the 30-tonne tank, negotiated a contract with the Hispano-Suiza company in 1961 to supply a suitable diesel engine to replace the SOFAM unit. The diesel offered superior low-end torque characteristics to a petrol engine in a battle tank, as well as far better fuel economy and lower risk of engine fire. The thoroughly impractical requirement issued by NATO in 1957 for the development of multi-fuel engines was also technically skirted by the adoption of a diesel engine. Hispano-Suiza built six HS-110 'multi-fuel' engine prototypes for AMX through the last months of 1962 and in early 1963. Five of these were intended to be mounted in AMX30 prototypes and AMX30As, while the fifth was kept as a source of spares and to reference any modifications.

In early 1963 engine number six was mounted in AMX prototype number 1 (W510-207) where it exceeded 300 running hours and over 3,000km without incident. The HS-110 was then evaluated directly against the SOFAM 12 GSDS petrol engine. The SOFAM had an output of 650hp at 2,750 rpm, whereas the HS-110 had an output of 720hp at 2,600rpm (governed to 690hp at 2,450rpm). Despite the difference in horsepower and the diesel's slower acceleration, the HS-110 demonstrated better low-end torque. At the end of 1963 the HS-110 was selected for the production vehicle and was heavily tested over the following three years. During this time DTAT facilitated a contract in December 1964 between Hispano-Suiza and SAVIEM to license manufacture the HS-110 at Limoges. In March 1965 the responsibility for the AMX30's engine was arranged as follows: AMX retained overall control of the engine programme, SAVIEM was the main manufacturing contractor, and Hispano-Suiza remained design parent with responsibility for technical development of the engine. A second contract for 24 more engines was placed immediately for the final development programme pursued with the AMX30As and for the two AMX30 *chars de définition* constructed in 1965.

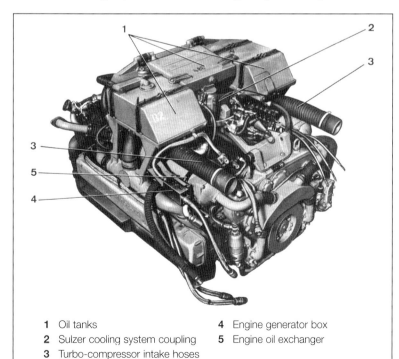

1	Oil tanks	**4**	Engine generator box
2	Sulzer cooling system coupling	**5**	Engine oil exchanger
3	Turbo-compressor intake hoses		

The 28.7-litre displacement HS-110 diesel was a flat 12-cylinder producing 710hp but governed to 680hp at 2,600rpm, generating 1,623ft/lb of torque at 2,000rpm. This gave a power-to-weight ratio of 18hp/tonne and a maximum speed of nearly 60km/h on a paved road. The production manual transmission was equipped with five forward and one reverse gear. The engine was liquid cooled, necessitating adequate room for the cooling system and radiators, with a displacement of 26.4 litres of coolant. No auxiliary engine was provided in the manner common to US or British practice. The AMX30B's turning radius was 3m in first gear, it could climb a gradient of 60%, could manoeuvre on a tilt of up to 17%, climb a 1m step and cross trenches up to 2.9m wide. Drivers soon became aware of the reserve of power available, but the clutch on early vehicles was a weak point that was addressed through careful training and eventually through the provision of more robust components.

In terms of reliability, the AMX30B came under criticism for transmission failures in its early years, most of which were as a result of driver error. Drivers previously qualified on the M47 often required a considerable education when conversion to the AMX30B began. Regular training of drivers led to a special driver-trainer version of the tank that could be implemented by means of a standardised kit which prescribed the removal of the turret and the installation of a caged-in passenger position. French cavalrymen and *tankistes* nicknamed this version '*La Gueuse*' (or vagabond) and examples could be found at any of the training centres responsible for training tank crews. Unburdened of their heavy turrets, the driver training tanks were faster and rode higher than standard AMX30Bs. The same kit was later widely adapted to the AMX30B2 and even to a surviving AMX32 hull.

A single carbon dioxide-type fire extinguisher bottle was installed inside the engine compartment in the event of fire. Just under 970 litres of fuel was carried in six fuel tanks in the hull, giving the AMX30B a road range of 644km under ideal conditions. Cross-country operation was of course far harder on fuel economy, but the AMX30B's fuel consumption was considered to be satisfactory. It would have

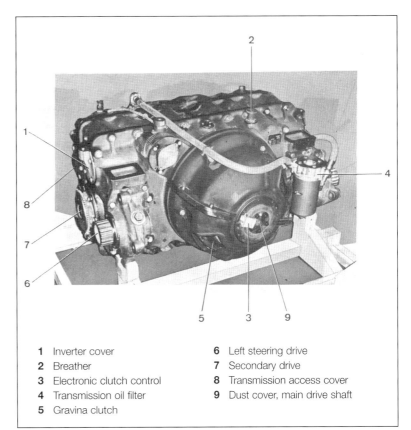

1 Inverter cover	**6** Left steering drive
2 Breather	**7** Secondary drive
3 Electronic clutch control	**8** Transmission access cover
4 Transmission oil filter	**9** Dust cover, main drive shaft
5 Gravina clutch	

ABOVE The BV5-SD-200D manual transmission was a five-speed manual unit designed by DEFA and it proved to be one of the weaker points of the overall AMX30B design. However, it posed fewer problems for an experienced driver given adequate maintenance and regular inspection after improved components were introduced. While clutch failures were common after 2,500km, the AMX30B was actually a more reliable vehicle than many of its contemporaries. *(Collection Thomas Seignon)*

looked positively stellar in comparison to the short-ranged, petrol-fuelled M47.

The AMX30's suspension system was referred to from the outset as the Vickers-type within DEFA. This suspension was selected due to its relative compactness and simplicity. It employed ten torsion bars, laid out alternating from the right-side units to left-side units across the length of the hull. Each torsion bar had a swing arm fitted with an axle, mounting a double-tyred road wheel. The drive was from twin sprockets mounted at the rear of the chassis, while the length of track was adjusted or tensioned through a front-mounted idler wheel assembly. Though the use of a torsion bar suspension represented nothing novel, the use of leading arms and trailing arms

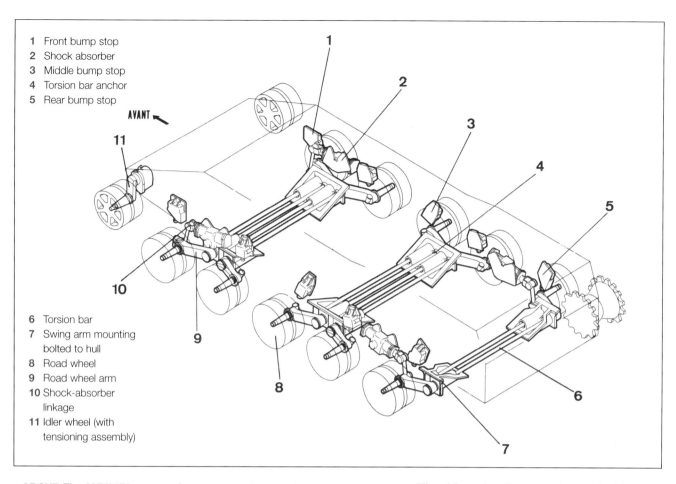

1 Front bump stop
2 Shock absorber
3 Middle bump stop
4 Torsion bar anchor
5 Rear bump stop

AVANT ←

6 Torsion bar
7 Swing arm mounting
   bolted to hull
8 Road wheel
9 Road wheel arm
10 Shock-absorber
   linkage
11 Idler wheel (with
   tensioning assembly)

ABOVE The AMX30B's suspension components were also manufactured at Roanne. Its running gear included a front-mounted idler and rear-mounted final drives and drive sprockets. The vehicle rode on five double road wheels on each side. The track was of the rubber-shod single-link 'dead' type; 0.57m wide, suspended on each side by five return rollers. It was tensioned by two large screws running through the front plate to the idler assembly on each side. The idler was adjusted to compensate for wear by means of a large set of ratchet-type wrenches. The width of the track and the low weight of the vehicle gave it a respectable ground pressure of just under 11psi at full combat loading, with a consequently good cross-country performance. The components, less the drive sprockets and return rollers, are illustrated above in simplified form. *(Collection Thomas Seignon)*

RIGHT The single-pin track developed for the AMX30B was made up of steel links with a central guide tooth. The rubber track shoe was attached from the inside face with two countersunk bolts.
(Pierre Delattre)

differed from American practice on the M26 and M47 (and from West German custom demonstrated on the *Standardpanzer*). The use of leading arms on the AMX design allowed a shorter wheelbase (and consequently a more compact hull). This proved an advantage in keeping overall weight down, but it brought the disadvantage of a rougher ride than might have been possible with a suspension made up of trailing arms.

The first, third and fifth road wheel stations were equipped with leading swing arms, and the second and fourth stations were fitted with trailing arms. Five return rollers supported the upper run of track. The first and fifth suspension unit on each side was equipped with an external shock absorber. Further development of this basic suspension system was conducted by installing heavier torsion bars and by fitting improved shock absorbers on derivatives including the AU F1, the AMX32 and the AMX30B2. Beyond these incremental improvements, the AMX30's suspension remained essentially unchanged between the

LEFT The idler wheel seen here on an AMX30B2 was the same for all versions of the AMX30. It was a double wheel, consisting of two cast spoked wheels, with rubber tyres. We can also see the bump stop for the No 1 road wheel station. *(Pierre Delattre)*

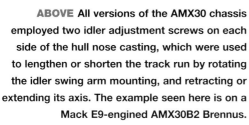

ABOVE All versions of the AMX30 chassis employed two idler adjustment screws on each side of the hull nose casting, which were used to lengthen or shorten the track run by rotating the idler swing arm mounting, and retracting or extending its axis. The example seen here is on a Mack E9-engined AMX30B2 Brennus.
(Pierre Delattre)

ABOVE The AMX30P Pluton prototype preserved at Saumur was fitted with standard AMX30 wheels, suspension units and drive sprockets. The cast road wheels each featured eight reinforcing ribs. The sprockets consisted of twin wheels with six spokes; the sprocket drive rings were replaceable and were attached from the outside with 16 bolts. The sprocket wheels were fixed to the final drives with 12 bolts. The Nos 1, 3 and 5 road wheel stations were carried on leading arms, while the Nos 2 and 4 stations were carried on trailing arms. The Nos 1 and 5 road wheel stations on each side were equipped with a shock absorber, seen here. Note that the AMX30P was equipped with a different exhaust system from the AMX30B battle tank.
(Pierre Delattre)

first AMX30B constructed in 1966 and the end of production of the EBG in the early 1990s. The limitations of the AMX30's basic chassis design led GIAT's AMX/APX design team in the early 1980s to come up with an entirely new chassis, suspension system and hull for the AMX40.

The road wheels, idlers and sprockets were all cast steel assemblies, produced and machined at ARE. The AMX30's suspension components proved very resilient and – with the exception of normal wear to the rubber components and bearings – proved very satisfactory in service. The design of individual components, like road wheels, idlers and sprockets, meant that these could be easily rebuilt. The original single-pin tracks themselves were good enough in basic

LEFT The AMX30B2 suspension saw detail improvements to the same general system developed for the AMX30B. It included thicker torsion bars, improved shock absorbers and in due course, double-pin tracks. These elements all evolved through the course of developing AMX30 variants (which were heavier than the original battle tank). The basic 'Vickers' suspension system (as it was known in France, probably because the general workings were based on existing Vickers patents when the AMX30 chassis was designed in the late 1950s) was retained for all AMX30 derivatives. The thicker torsion bars and improved shock absorbers gave a more comfortable ride and made the AMX30B2 a more stable gun platform.
(Pierre Delattre)

RIGHT The upper run of tracks was suspended by five return rollers, which supported the inner half of the links, stopping at the guide pin. *(Zurich 2RD)*

RIGHT The No 5 right-hand suspension unit on a preserved AMX30B, showing the accompanying torsion bar mounting to its immediate left. *(Zurich 2RD)*

ABOVE The layout of the torsion bars through the AMX30B's hull floor resulted in an asymmetric layout on each side of the hull. The left- and right-side torsion bars were positioned to alternate closely. Each torsion bar mounting base was anchored through the hull side with six bolts. Upgrading the suspension with larger-diameter torsion bars was a simple matter given this layout, but was also limited by the tight clearances necessary. *(Zurich 2RD)*

design to remain in service throughout the AMX30B's life and to have been used on later designs extending up to the Leclerc prototypes.

Removing worn tracks and fitting a new set was of course a massive undertaking for any tank crew in any army. The tank was driven to the flattest ground possible, and an area of perhaps 100m was ideal, preferably a peacetime tank park, for rougher ground often required the assistance of a recovery vehicle.

The AMX30B's road wheels were chocked,

LEFT The *Chenilles à Connecteurs* (double-pin track system) were introduced for the AMX30 family in the late 1980s, although the first tracks of this type had been trialled in 1973. Their introduction came as a result of testing alternate tracks suitable to higher-speed vehicles, and were seen on the AMX32 and AMX40 for export. By the late 1990s the double-pin tracks were widely used on AMX30 variants like the AMX30R Roland 2 and the AU F1 T, as well as on the AMX30B2. They were seldom, if ever, used on the AMX30B. *(Zurich 2RD)*

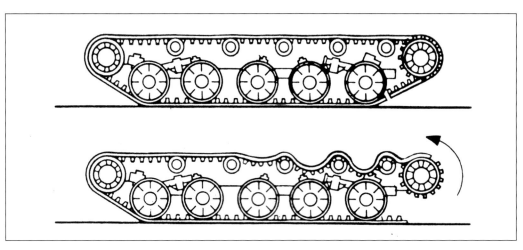

LEFT The prescribed method of track removal is shown here. *(Collection Thomas Seignon)*

and the worn track would be 'cut' by knocking out a track pin between the drive sprocket and the No 5 road wheel station. The driver would set the gearbox to neutral and the crew would slowly tug the track across the return rollers from the idler, laying out the length as they went, until the whole track was laid out on the ground in front of the tank. The first and fifth road wheels would be chocked and the procedure repeated for the opposite side.

Reinstalling new track was an even more difficult business. The new track would be assembled and laid out ahead of each side of the tank, and the AMX30 might be towed slowly on to the new run of tracks by a recovery vehicle (or another tank). A wire rope was attached from the last link to the drive sprocket, and the driver put the transmission in gear, slowly winding the top run across the return rollers and back towards the sprocket. The wire rope was removed, and the track was levered to join just behind the fifth road wheel and the last track pin was knocked in. The procedure was repeated on the other side. This all sounds simple enough, but the reality was a day of back-breaking labour in hot sun or bitter cold.

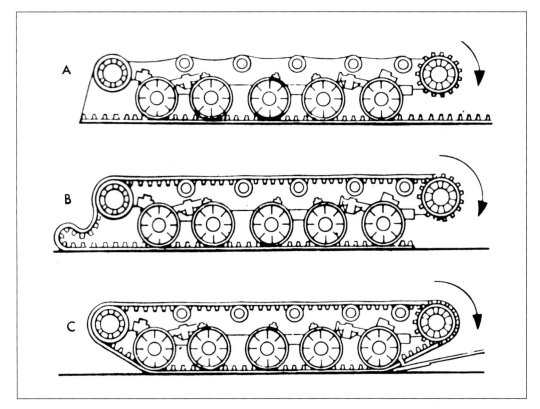

LEFT Reinstallation of the track is demonstrated in this diagram, which belies the huge physical effort involved for the crew. *(Collection Thomas Seignon)*

RIGHT AMX30B cutaway. *(Ian Moores)*

1 Rigid thermal sleeve (4 section)
2 Muzzle, CN-20-F2 20mm automatic cannon
3 Rubber waterproof 20mm barrel sleeve
4 Dust cover
5 20mm cannon mantlet sleeve
6 Main mantlet and mantlet dust cover assembly
7 Right side turret exterior stowage basket
8 Coaxial armament mantlet (capable of de-coupled independent elevation)
9 M208 Rangefinder head, right-hand side
10 Brush guard for cupola machine gun
11 PH8B searchlight mounting
12 Gunner's M267 periscope
13 7.62mm cupola machine gun
14 Commander's PH9A searchlight
15 Commander's panel
16 Commander's M267 sight with M270 day prism fitted
17 Loader's M267 periscope
18 Circular cupola 7.62mm ammunition feed box.
19 Antenna
20 Antenna base
21 TOP 7 cupola (incorporates 10 x M268 periscopes)
22 M223 episcope
23 Gun cleaning rod stowage

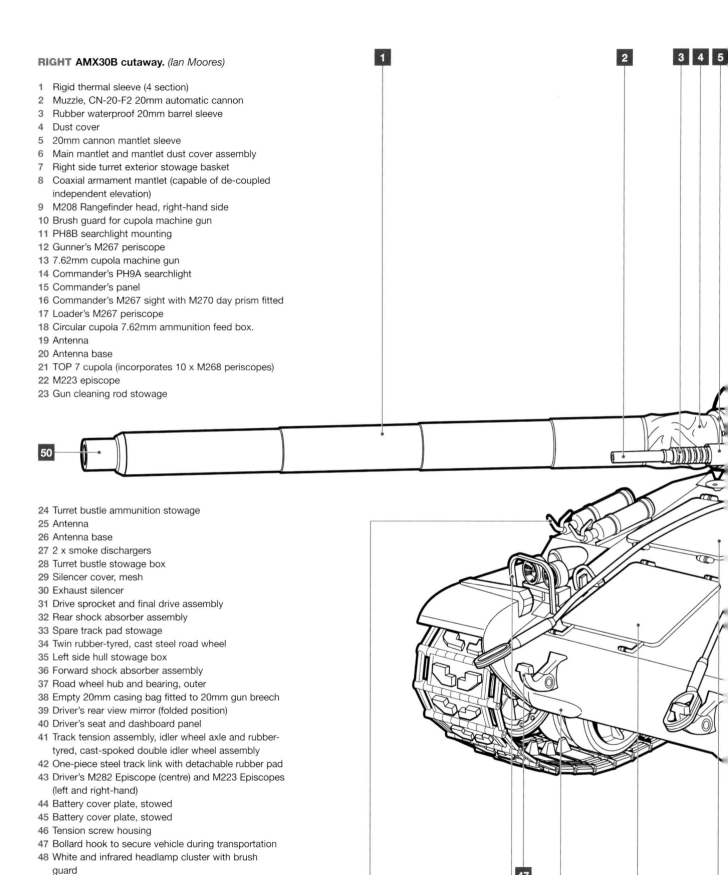

24 Turret bustle ammunition stowage
25 Antenna
26 Antenna base
27 2 x smoke dischargers
28 Turret bustle stowage box
29 Silencer cover, mesh
30 Exhaust silencer
31 Drive sprocket and final drive assembly
32 Rear shock absorber assembly
33 Spare track pad stowage
34 Twin rubber-tyred, cast steel road wheel
35 Left side hull stowage box
36 Forward shock absorber assembly
37 Road wheel hub and bearing, outer
38 Empty 20mm casing bag fitted to 20mm gun breech
39 Driver's rear view mirror (folded position)
40 Driver's seat and dashboard panel
41 Track tension assembly, idler wheel axle and rubber-tyred, cast-spoked double idler wheel assembly
42 One-piece steel track link with detachable rubber pad
43 Driver's M282 Episcope (centre) and M223 Episcopes (left and right-hand)
44 Battery cover plate, stowed
45 Battery cover plate, stowed
46 Tension screw housing
47 Bollard hook to secure vehicle during transportation
48 White and infrared headlamp cluster with brush guard
49 Fire extingulshers
50 CN-105-F1 105mm gun muzzle

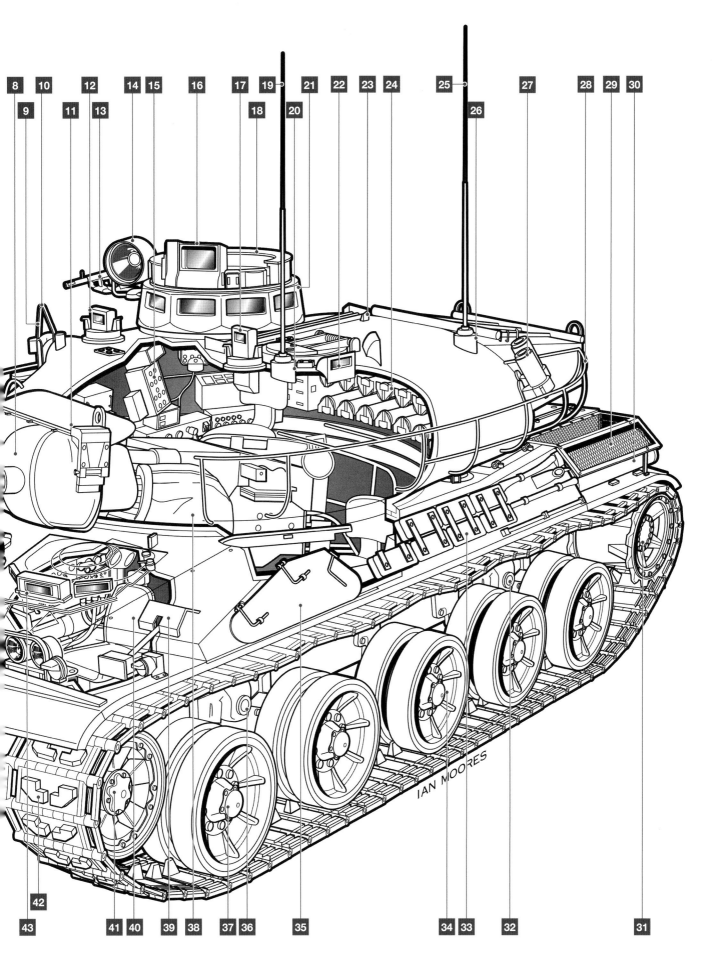

8
9
10
11
12
13
14
15
16
17
18
19
20
21
22
23
24
25
26
27
28
29
30

42
43
41
40
39
38
37
36
35
34
33
32
31

IAN MOORES

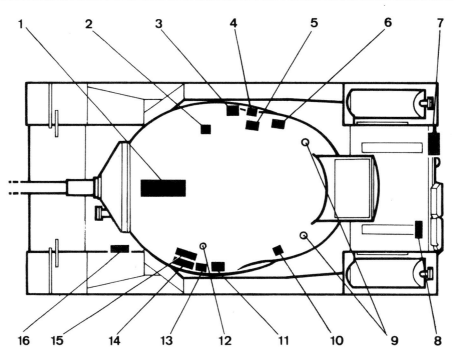

1. TR-VP-6-A or TR-VP-5-A radio set (early production tanks) or TR-VP-113-A radio set.
2. Gunner's radio/intercom command BC-220A or BC-320A
3. Commander's radio/intercom command BC-220A or BC-320A
4. FA-2-A or FA-5-A command link (to exterior telephone AT-17-A)
5. BC-270-A command panel for transmitter/receiver ER-100-A or BC-344-A
6. BC-270-A command panel for receiver RR-85A or BC-344-A
7. AT-17-A exterior telephone
8. BJ-197-A or BJ-217-A junction box
9. Antenna base
10. Hull radio-intercom junction box BJ-196-A or BJ-218-A
11. Loader/operator's radio/intercom command BC-220A or BC-320A
12. Antenna base
13. Radio-intercom junction box BJ-182-A or BJ-216-A (or BJ-232-A with transistor filter FI90-A)
14. AM-84 amplifier
15. Radio set TR-VP-13-A
16. Command box BC-169-A (or BC-293-A)

Communications

The AMX30B's radio equipment enabled the crew to communicate within the platoon and within the squadron radio net. Squadrons normally communicated with each other through their respective command tanks, but tanks could contact each other on designated frequencies. The radio equipment also served as the tank's intercom and junction boxes were located throughout the turret and the front part of the hull. The loader normally looked after the TR-VP-13-A set, while the commander's radio set varied in type depending on whether the tank was a platoon tank or a command tank. There were some changes introduced to the AMX30B's standard radio equipment during production. The first 50 AMX30Bs (built in 1966–67) were equipped with radio sets employing leads fitted with gland seal connectors. Two separate radio

fits were provided for these vehicles. The basic platoon vehicle was equipped with one TR-VP-5-A set and one TR-VP-13-A set. The command vehicle was equipped with a longer-range set TR-VP-6-A, an AM-84-C low-frequency amplifier and one TR-VP-13-A set. A change was made after the 50th AMX30B was built, retaining the same radio equipment types but using plug-in connections instead of gland seals.

From the 261st AMX30B built the TR-VP-13-A seat was retained but the TR-VP-5-A and TR-VP-6-A sets were replaced with the TR-VP-113-A set. At the same time the radio fits were changed to incorporate three standards: the platoon tank, the platoon commander's tank and the '*Char Colonel*', a command tank with an additional TR-VP-113-A set fitted. Bearing in mind that the radio equipment was spread throughout the AMX30B's turret, when the AMX30B2 upgrade was designed in 1979, changes had to be made

LEFT AMX30A No 5 (234-0289) photographed fording a shallow river. The fording capacity of the AMX30A, without snorkel or preparation for deep fording, was 2.20m (and remained so for the AMX30B). *(Fonds Claude Dubarry, Collection du Musée des Blindées de Saumur)*

(although the TR-VP-13-A and TR-VP-113-A sets were retained as standard equipment). Many of the locations previously employed for diverse elements of the radio systems were usurped by the 'black boxes' – the extensive suite of electronic components required for the COTAC system. As a result, the radio equipment was relocated into the turret bustle behind the commander, causing the NBC system to be relocated on to the exterior of the turret bustle.

Water-crossing capability and doctrine

Sealing the AMX30B's engine compartment and the rest of the hull to a waterproof standard for fording was specified from the early stages of the AMX30's design. Deep fording was tested and perfected with the AMX30A first production batch. The engine deck layout adopted for the AMX30A had even included permanently mounted shutter plates over the battery compartments, which could be slid closed prior to deep fording. The AMX30B could wade to a depth of 1.98m without preparation, but it required the inflation of the turret ring seal and the sealing of the battery compartment (among a host of other measures relating to the turret openings) in order to ford to full depth of 3.96m. Plates to seal the battery compartment were carried on the front glacis and preparing the vehicle took about 15 minutes. Two types of snorkel apparatus were employed for deep fording, to provide air to the engine as well as to the crew. The first of these, used strictly for training, was anchored over the loader's hatch

BELOW To prepare the tank for a peacetime or training water crossing, the crew required approximately 15–30 minutes. The crew would detach the battery compartment covers from the right side of the glacis, and clamp them over the battery compartment ventilation grilles. The exhaust flappers were released and the firewall aspiration was pulled open. The rangefinder covers were closed, the heater intake was plugged, the NBC vent was sealed, the shell ejection port was fastened and the fume extraction vent was plugged. An empty cartridge was chambered in each weapon to seal the breech. The 105mm gun was then raised to maximum elevation and its muzzle plug and cover was fitted. The cupola machine gun was dismounted and secured inside the turret. The coaxial weapon port was plugged in the case of the 12.7mm machine gun, or secured with a rubber muzzle cover in the case of the 20mm CN-20-F2. The crew then fitted the air intake tower to the open loader's hatch rim, securing the hatch open and stabilising the tower carefully with wire rope anchors. *(Fonds Claude Dubarry, Collection du Musée des Blindées de Saumur)*

opening, and was large enough for a crewman to climb through. A narrower type, intended to be used in combat only, was fitted to the loader's periscope aperture and could be carried stowed on the hull rear plate.

Submerged river crossing for the 30-tonne tank was a truly novel concept in French tank design in the early 1960s. It became an important element in French water-crossing doctrine, and an important reason why AMX30-based bridge-layer equipment was never procured by the army. The process of underwater crossing became a rite of passage for most AMX30B crews, practised often enough to merit permanent training facilities. Many of the integral waterproof features incorporated in the AMX30's design also functioned to seal the tank against NBC agents. The AMX30B was designed to draw air through a panel in its engine firewall when the engine compartment was sealed, aspirating through the turret. The exhaust pipes were fitted with spring-loaded flappers in order to function as one-way valves without risk of water ingress.

ABOVE Finally, the crew mounted the tank, the turret crew passing through the TOP7 cupola, first the loader, followed by the gunner and finally the commander. The engine was fired up, and the rotary junction seals were then inflated by the onboard air compressor. The turret lock was engaged and the traverse and elevation safety switches were engaged, for traverse or changes in gun elevation would rupture the inflatable seals. Now the driver proceeded carefully to a pre-surveyed crossing point under the commander's guidance. The presence of tree limbs had to be carefully noted and negotiated. Finally, with an AMX30D's winch rope ready for quick attachment by the ever-present engineer divers, the driver closed his hatch. The loader and commander exchanged places and secured the cupola hatch. The commander then climbed the rungs inside the training tower so that he could look out of the top, and the tank made its way to the river bank. (*Fonds Claude Dubarry, Collection du Musée des Blindées de Saumur*)

BELOW Listening carefully for any directions over the radio set and using the vehicle intercom, the commander carefully guided his driver through the murky water in a straight line to drive downwards and directly across the river bottom, and then directly up the opposite bank. (*Fonds Claude Dubarry, Collection du Musée des Blindées de Saumur*)

BELOW The much smaller combat snorkel, fitted here, was a much less forgiving piece of equipment, because it could not be used as an escape device. When using the combat snorkel, the crewmen were provided with a breathing device and were supported more extensively by combat divers, who could pull men out of a drowned tank. (*Fonds Claude Dubarry, Collection du Musée des Blindées de Saumur*)

Every hatch and port on the entire tank were provided with a watertight rubber seal, and every rotary joint was supplied with an inflatable pneumatic seal.

The waterproofing process was to permit the AMX30 to move across water obstacles up to 4m deep and was not intended for use in opposed crossings. In training, the vehicle commander would control the crossing from the top of the training tower, instructing the driver through the AMX30's intercom. In a combat situation, the entire process was guided by radio from the shore. Within a few minutes of emerging from the water, the tank could be readied for combat, a process which required the removal of the snorkel and replacement of the loader's periscope, the removal of the battery compartment covers and the closure of the firewall intake. The AMX32 carried forward the same capability. The process was almost identical in the AMX30B2 (and the AMX30B2 Brennus). It was similarly followed by the AMX30D, which was issued to armoured regiments and was expected to have the same mobility. The AU F1 was also designed for deep wading and could wade almost to turret roof depth – but no other AMX30 variants could ford deeper than 2m.

The NBC system

The NBC filtration system provided for the AMX30B was carried inside the turret, aspiring through a vent intake located on the floor of the turret bustle. The system was quite

ABOVE LEFT Special test basins were built at Satory and at Bourges among other places, all to test the deep-fording capabilities of AMX30 series vehicles. *(Fonds Claude Dubarry, Collection du Musée des Blindées de Saumur)*

ABOVE The final mission of AMX30A 234-0287 was to serve as a *Char Gloutte* – a specialised training vehicle employed for basic deep-fording training. The stripped and trackless 234-0287 was winched across a sloped-bottom concrete basin to get newly conscripted crews ready for the ordeal of submerged river crossing. *(Fonds Claude Dubarry, Collection du Musée des Blindées de Saumur)*

BELOW The NBC system employed on the AMX30B fit neatly behind the commander's position and was controlled from his panel. Each crew member was also equipped with a standard gas mask for use if the gun had to be fired or if any of the hatches were opened in a contaminated area. Here is the unit seen from the commander's side. Air was sucked in by a centrifugal fan mounted in the assembly on the left-hand side of the unit. *(Collection Thomas Seignon)*

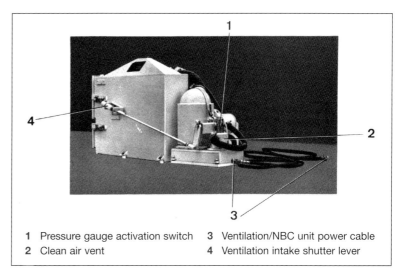

1 Pressure gauge activation switch   3 Ventilation/NBC unit power cable
2 Clean air vent   4 Ventilation intake shutter lever

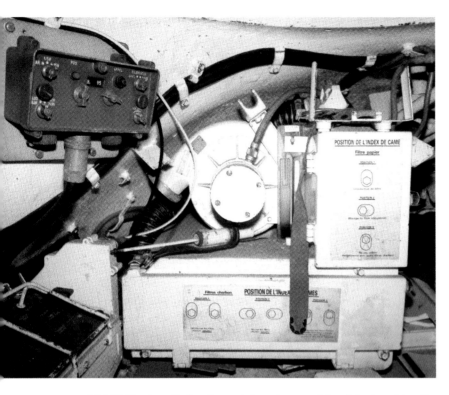

ABOVE The installation fit neatly into the rear right of the turret bustle.
(Thomas Seignon)

BELOW This is the unit's filter caisson. The large box on the bottom of the picture held two charcoal F1 filters for the absorption of vapours, the upper right box held a treated paper F1 filter. *(Collection Thomas Seignon)*

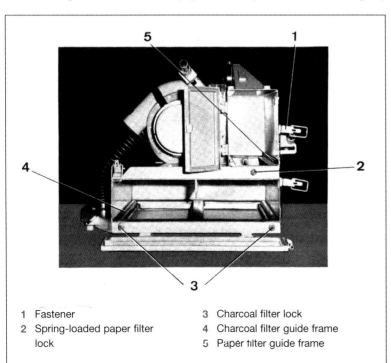

1 Fastener	3 Charcoal filter lock
2 Spring-loaded paper filter lock	4 Charcoal filter guide frame
	5 Paper filter guide frame

compact and was carried immediately behind the commander's position on the right-hand side of the turret bustle. It employed a centrifugal fan unit to draw outside air through a special 250m^3/h F1 paper (for dust) and then two 90m^3/h F1 charcoal vapour filters, pressurising the tank's interior with filtered air. The NBC system was controlled from the commander's station and featured a central pressure gauge, which was calibrated with hatches open, establishing a baseline atmospheric pressure maintained by the ventilator unit. The pressure gauge maintained the overpressure level and controlled the fan's operation.

To use the system, the commander activated the NBC system switch from his control panel, then switched the operating mode from the 'ventilation' to the 'NBC' position. All hatches and covers were closed, the crew heater air intake was sealed and the fighting compartment firewall air intake to the engine was closed. A round (or empty casing) was loaded into the 105mm gun and the secondary armament. Adjusting the pressure dial to match the liquid level in the pressure gauge controlled the air aspiration, which created a positive air pressure level inside the tank.

Had the system been used in a contaminated area, the tank and all its hatch openings had to be decontaminated from the outside prior to switching the NBC system back into ventilator mode again, and the filters had to be changed. This process normally required a full spray down with a water/detergent mixture. The original NBC system was replaced by a modernised system located on the exterior of the turret (with filter elements changed from the exterior of the turret) when the AMX30B2's T105M turret was developed and the radios moved into the turret bustle. While operation was essentially the same, the updated system's filter box could no longer accommodate two 750ml bottles (typically champagne), which would be cooled to ideal drinking temperature by the operation of the system in the AMX30B. Maintenance of the NBC system was a task undertaken at regimental or divisional workshops.

OPPOSITE An AMX30B2 being decontaminated in the 1990s. *(Peter Lau)*

ABOVE It was vital at the conclusion of any operation with the AMX30 series tanks in a contaminated area to spray down all equipment with water and detergent. While the scene shown here is a decontamination procedure exercise under way over a decade after the Cold War involving an EBG and an AU F1 TA, it gives a very good idea of what decontamination entailed.

(Pierre Delattre)

ABOVE The NBC caisson fitted to the rear of the AMX30B2's turret included an improved filtration unit whose filters could only be changed from the outside of the tank. This took the place of the smaller turret bustle stowage box carried by the **AMX30B.** *(Zurich 2RD)*

Chapter Three

Main armament and sighting equipment

The DEFA 105mm gun, unlike the British Royal Ordnance 105mm L7, had already been under development for years by the time of the 1956 Hungarian uprising. Refined into the D1512 and adopted as the CN-105-F1 for the AMX30 in 1962, the French gun went on to outperform its rival at the Mailly trial. Provided with high-velocity kinetic energy ammunition, the CN-105-F1 was an accurate, reliable and well-engineered weapon.

OPPOSITE The CN-105-F1 105mm gun stood virtually unchanged from 1962 until the present day (this example was seen on an AMX30B2 in 2006). A well-designed and extremely well-made gun, it proved capable of firing high-velocity kinetic energy rounds as comfortably as the lower-velocity Obus-G hollow-charge round. The main modifications entailed were made to the sights rather than to the gun itself. *(Pierre Delattre)*

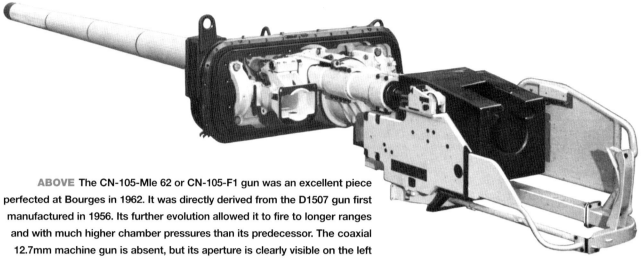

ABOVE The CN-105-Mle 62 or CN-105-F1 gun was an excellent piece perfected at Bourges in 1962. It was directly derived from the D1507 gun first manufactured in 1956. Its further evolution allowed it to fire to longer ranges and with much higher chamber pressures than its predecessor. The coaxial 12.7mm machine gun is absent, but its aperture is clearly visible on the left side of the mount. The gun's recoil mechanism incorporated two diametrically opposed hydraulic cylinders which functioned as brakes. An oleopneumatic recuperator ran the gun back into battery. The breech was of the semi-automatic type. The gun was loaded from the left-hand side. In the 1970s, armour-piercing fin-stabilised discarding sabot ammunition was developed to fire from the unchanged CN-105-F1, which was adopted for the AMX30B2. The CN-105-F1 was also chosen to arm the first AMX32 prototype turret. (Nexter)

The French 105mm gun was a weapon with a history that stretched back to the study of captured German tank guns at the end of the war. On 1 June 1946 an evaluation project began at ARE to evaluate the components of the PAK 43/41 88mm gun, which was by French estimations the best German tank gun fielded in the recently ended conflict. Ironically the British L7 105mm also traced its lineage in part to this same weapon, via the 20pdr. The breech system employed in the design of the PAK 43/41 was adopted in DEFA's early 105mm gun designs perfected in the mid-1950s. These emerged in the DEFA 105mm tank gun designated D1507, whose early development had been pursued at the *Section Technique d'Artillerie* at St Cloud, and then perfected at

the *Arsenal de Bourges*. The D1507, which followed the principles of the PAK 43/41's design (with the exception of its rifling, which we will see was due to its specific ammunition types) was developed into the CN-105-57. The CN-105-57, firing hollow-charge HEAT ammunition, outshone the contemporary 90mm high-velocity gun firing kinetic ammunition.

In France HEAT ammunition (and the Obus-G in particular) seemed the long-term solution to defeating heavy Soviet armour – as well as the doctrinal means of projecting heavy firepower from relatively light AFVs. French design practice oriented itself around the model of the lightest possible vehicle deploying the heaviest possible firepower and the best possible mobility. The rationale of building AFVs (specifically battle tanks) with heavy armour ran counter to this new doctrine, and was set aside. Alongside other chemical energy anti-tank weapons, the Obus-G seemed to spell the end of heavy armour. The doctrinal impact of the chemical energy projectile deeply affected the design of DEFA's 105mm guns over the following two decades. In particular, the CN-105-57 and CN-105-F1's rifling employed a less aggressive twist in comparison to that of the British L7 105mm, in order to better fire the Obus-G to optimal range.

Perfecting the 105mm gun from the CN-105-57 into the D1512 gun for the 30-tonne medium tank followed in the period 1957–65. The long development time was in part due to the fact that a complete set of sighting equipment was designed alongside it, intended for use by day and by night. The D1508,

RIGHT A look through the breech down the barrel of a CN-105-F1 105mm gun. (Zurich 2RD)

D1509, D1510 and D1511 DTAT gun designs were all related versions of the same 56-calibre-length 105mm guns.

Each had detail changes in rifling or other construction details specific to firing different experimental ammunition types or in order to explore the possibility of developing a respectable armour-piercing discarding sabot (APDS) KE round. These guns shared specific requirements for higher chamber pressure ammunition than the CN-105-57 to impart optimum range. The D1512 featured a stronger breech and chamber, a longer barrel and a pneumatic bore evacuation system. The decision not to use a muzzle brake on the D1512 came as a result of extensive testing with the CN-105-57 and of the earlier high-pressure guns, which proved that it was much easier for the gunner to follow the shell's tracer element without such a device.

The D1512 was adopted in 1962 as the Canon 105mm F1 (CN-105-F1) as the chosen armament for the new 30-tonne medium tank. The gun was also officially known in its early days (much less commonly) as the CN-105-Mle 62 in reference to the year in which it was officially adopted. The designation referred to the complete 56-calibre-length gun. When the CN-105-F1 gun specification was established at the *Arsenal de Bourges* it specified metallurgical requirements for the use of 30 NCD 12 nickel-chrome-molybdenum steel of the highest quality to impart exceptional strength to the gun. The CN-105-F1's specification went so far as to specify that only the product from CCNM steel mill at Montlucon could be used for barrel construction and named Aubert et Duval as approved suppliers for steel for the gun's breech components. Other producers were employed in sourcing materials for the CN-105-F1 gun, but this measure was adopted to forewarn the *Arsenal de Bourges'* network of suppliers that strict adherence to quality standards would be checked regularly, perhaps for the first time. As a result, the CN-105-F1 was an exceptionally durable piece and was built stoutly enough to be seriously evaluated for re-boring into 120mm calibre in the 1980s (although this option was never pursued).

In 1965 the CN-105-F1 gun entered production at the *Arsenal de Bourges*, and it underwent very little change for the remainder of its service life in the AMX30B or the AMX30B2. The CN-105-F1 was fired by means of an electrical firing circuit in all of its prototype forms and in the case of the production gun. In its production form it measured 5,900mm overall, with a 5,263mm rifled bore and a chamber length of 637mm. Its recoil travel was 400mm, exerting a recoil force of 18 tonnes. It fired the OCC-61 round at a velocity of 1,000m/sec. Fume extraction was achieved by compressed air injection into the bore immediately after firing. This permitted the rapid evacuation of any fumes that might have entered the turret upon the breech being opened, and compressed air

was supplied from the fixed tank mounted on the hull floor next to the driver.

The 30-tonne tank's main armament programme did not ignore foreign innovations in gun design, and the eight-year development time between 1957 and 1965 included side projects to validate the choice of the Obus-G and other design parameters. It was decided to auto-frettage the bores on several of the early prototype D1512 guns, and this process was adopted for the production CN-105-F1. The evaluation of rigid thermal sleeves for the CN-105-F1 was conducted throughout 1964. The rigid type of sleeve cooled far more rapidly than the flexible blanket-type thermal sleeve,

BELOW The top of the CN-105-F1's breech, again in an AMX30B2. The breech occupied much of the turret's internal volume. Each gun was marked 'D1512' and carried a serial and manufacturer's markings from Bourges, France's greatest artillery arsenal. *(Zurich 2RD)*

while both types functioned well in decreasing 'barrel bend' and contributed to higher accuracy at maximum ranges. After this was established with data gathered in comparative DTAT firing tests between two AMX30As equipped with the different sleeve types on 19 June 1964, the rigid type was adopted for the CN-105-F1 production guns.

The British innovation of using a two-piece bagged charge ammunition type was also investigated for the DEFA 105mm gun, to the extent that a test gun was built using a special breech sealed with a set of obturator plates. This gun underwent firing trials in 1964 at Bourges, mounted on a firing stand for full comparison to one of the D1512 prototypes firing fixed ammunition. While this breech arrangement was not pursued further for the CN-105-F1, the bagged charge version of the gun worked very well and had an identical performance to its conventional version. DTAT's tests on the basis of the obturator breech type fitted to a D1512's barrel proved vital to the eventual development of the 155mm GCT gun breech employed on the AU F1 self-propelled gun in the mid-1970s.

Secondary armament – 12.7mm and CN-20-F2

The AMX30B was ordered into production with a specified coaxial armament consisting of a licence-built M2 Browning 12.7mm heavy machine gun. The intention was to use the heavy machine gun as both an anti-personnel weapon, against anti-tank guns and against softer targets like trucks and APCs. Moreover, the eventual adoption of an even heavier 20mm secondary armament that could be employed against non-MBT targets at ranges up to 1,000m was specified as an additional requirement at around the time the AMX30B entered production.

The 20mm M693 gun's development path stretched back to another FINABEL prerequisite of 1957. The requirement to improve the AMX30B's secondary armament matured as the DTAT M693 20mm cannon design, which went on to arm the AMX10P and AMX30 as the CN-20-F1 and CN-20-F2 respectively. The CN-20-F2 was a belt-fed automatic weapon which fired a 213mm-long 20mm round derived

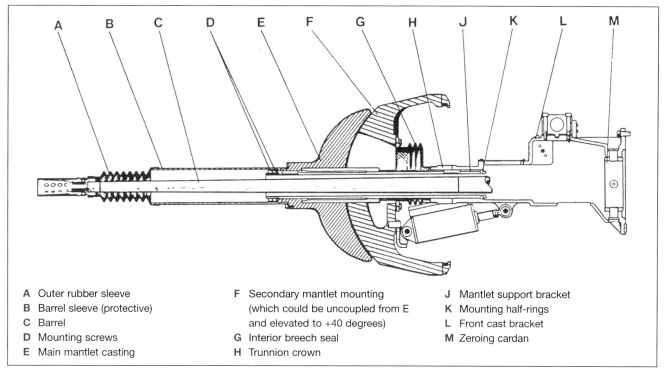

A Outer rubber sleeve
B Barrel sleeve (protective)
C Barrel
D Mounting screws
E Main mantlet casting

F Secondary mantlet mounting
(which could be uncoupled from E
and elevated to +40 degrees)
G Interior breech seal
H Trunnion crown

J Mantlet support bracket
K Mounting half-rings
L Front cast bracket
M Zeroing cardan

from the Swiss HS820, with a muzzle velocity of 1,050m/sec at a maximum rate of fire of 750 rounds per minute. The 20mm cannon was ready in 1971 but the programme for the AMX10P was given higher priority and as a result the AMX30B's field trials with the new weapon took place in early 1974. By February 1974 the weapon was approved and in 1975 the first CN-20-F2s were installed in new-build AMX30Bs, while a retrofit programme was put into place to modify existing AMX30Bs

ABOVE A simplified view of the 20mm CN-20-F2 mounting seen on late and rebuilt AMX30Bs, the AMX32 and on all AMX30B2s. Note the deep fording muzzle cap is not shown, and the breech detail is missing. *(Collection Thomas Seignon)*

during base rebuilds. The retrofit required the installation of the 20mm cannon into a modified mantlet as well as the installation of the 20mm feed system inside the turret. A total of 480 rounds of 20mm belted ammunition were normally carried.

RIGHT A late-production or rebuilt AMX30B fitted with the CN-20-F2 (M621) 20mm cannon. The inside left part of the turret has been modified to fit the 20mm cannon and to hold 400 × 20mm rounds. The modification was implemented without hampering the maximum elevation of the weapon, which was still able to be uncoupled from the main gun and elevated to +40 degrees. Illustration of the maximum elevation that can be reached by the main gun (+20 degrees) and by the coaxial weapon (+40 degrees), here the 20mm cannon. The decoupling of the coaxial weapon was intended to provide an anti-aircraft capability against slow-flying aeroplanes or helicopters. *(Fonds Claude Dubarry, Collection du Musée des Blindées de Saumur)*

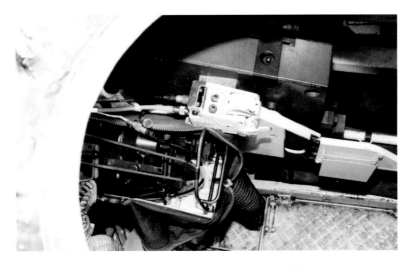

LEFT Inside view of the turret seen from the loader's hatch. The rear of the 20mm cannon's breech is clearly visible in its unzipped protective canvas breech cover. The little white box with the green and red lights is part of the safety equipment for the main gun-firing sequence: when a new 105mm round was chambered, the loader had to push on the large contact breaker to enable firing by the gunner. He would shout 'ready' at the same time to indicate the gunner could fire. *(STAT)*

The CN-20-F2's 20mm HE round weighed 310g and included a 120g projectile containing 10g of explosive. The 20mm armour-piercing round could penetrate 20mm-armour plate inclined at 60 degrees at over 1,000m distance, a fine performance that anticipated the BMP1's introduction in a timely manner. The gun was normally locked to the same elevation as the main armament, but it could also be elevated independently of the CN-105-F1 to permit high-angle fire against helicopters, aimed by the commander through the TOP7 cupola's sight.

The 20mm secondary armament was very successful in the AMX30B and was retained for the AMX32, AMX30B2 and even for the AMX40. The only drawback of the CN-20-F2 was that it occupied a larger amount of space inside the turret than its predecessor, and its feed system had to be carefully inspected daily. The feed system was vulnerable to damage,

CENTRE The CN-20-F2 was equipped with a large ammunition feed system, which took up considerable space inside the turret to the left of the main armament. This example is fitted inside an AMX30B2, but a substantial number of the AMX30B production run was refitted with the CN-20-F2 during base rebuilds or even in workshops. *(Zurich 2RD)*

LEFT The CN-20-F2's ammunition was substantially larger than the 12.7mm machine-gun's rounds, but the breech of the weapon itself did not intrude as far inside the turret as many other weapons of this type might have. Used in action during Operation Daguet, the CN-20-F2 was ideal for engaging light AFVs. *(Zurich 2RD)*

which could deform the steel trackways and cause stoppages. Not all AMX30Bs received the CN-20-F2 and at the end of the Cold War a considerable number of AMX30Bs (including most of those issued to the mechanised infantry regiments) retained the 12.7mm coaxial machine gun. Empty 20mm ammunition boxes were generously sized and became one of the few accepted forms of extemporised stowage for the AMX30 series tanks – in an army which prided itself on careful conformity to the rule book where stowage was concerned!

The Obus-G HEAT round

DEFA explored the development of hollow-charge high-explosive anti-tank ammunition in the 1950s. A sophisticated hollow-charge round based on German wartime research was designed in the early 1950s at the Institut Franco-Allemand de St Louis by the German engineer Gessner. Spinning a hollow-charge round through a rifled gun barrel to impart accuracy degraded the effect of its explosion on impact. Gessner's response to this was to enclose the HEAT hollow charge –

suspended on a ball bearing race to minimise spin – within an outer streamlined shell. The result was a HEAT round with optimised armour penetration, which could be fired accurately to a longer range than other HEAT round configurations. The French described the Gessner round as the Obus-G. The Obus-G had to be manufactured to very precise specifications and thus cost much more than simpler types of HEAT rounds.

The complete 105mm Obus-G round weighed 22.2kg, with a projectile weight of 10.95kg. It was fired at a velocity of 1,000m/sec and showed a dispersion of 1.5 milemes at 1,500m, which was consistently more accurate than 105mm APDS rounds available in the early 1960s. It could penetrate 150mm of rolled homogenous armour sloped at 65 degrees at all combat ranges, or 360mm of vertical armour plate. The Obus-G was a relatively complex and costly round to produce, and it could only be adopted for one calibre for budgetary reasons. This was standardised for the French Army in 1961 as the 105mm OCC-105-61.

The DEFA-sanctioned belief in the primacy of

BELOW The CN-105-G1's ammunition: from left to right we have the 105mm OE-105-F1 (high-explosive), high-explosive anti-tank (OCC-105-Mle 62, or OCC-105-F1, alias Obus G) and OEC-105-F1 Illuminating rounds with typical markings. *(Collection Thomas Seignon)*

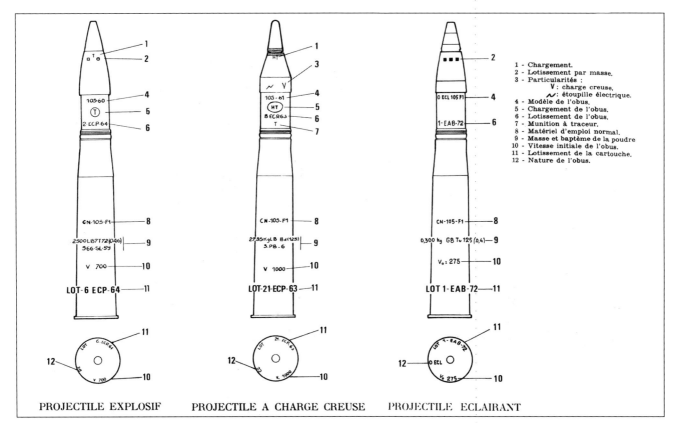

1 - Chargement.
2 - Lotissement par masse.
3 - Particularités :
    V: charge creuse,
    ∿: étoupille électrique.
4 - Modèle de l'obus.
5 - Chargement de l'obus.
6 - Lotissement de l'obus.
7 - Munition à traceur.
8 - Matériel d'emploi normal.
9 - Masse et baptême de la poudre.
10 - Vitesse initiale de l'obus.
11 - Lotissement de la cartouche.
12 - Nature de l'obus.

PROJECTILE EXPLOSIF   PROJECTILE A CHARGE CREUSE   PROJECTILE ECLAIRANT

HEAT anti-tank rounds set the path for French thinking on how best to destroy enemy armour up until the aftermath of the Yom Kippur War. In a sort of chicken-and-egg manner, the belief in the primacy of the HEAT round also affected how the French determined their own AFVs' needs in terms of armoured protection. If after all heavy armour was so vulnerable to weapons like the Obus-G, it seemed altogether pointless to develop tanks with heavy armour. Tanks would of course need to be frontally immune to small-calibre fire and to artillery barrages, but there was no point in encumbering the 30-tonne tank with excess weight when the

vehicle's very mobility was understood to provide much of its overall protection.

The adoption of the Obus-G did not mean that KE armour-piercing rounds were ignored by DEFA's research teams in the late 1950s, but rather that HEAT was better suited to French capabilities and tactical concepts. The French Army and DEFA (and its successors) heavily backed the development of hollow-charge munitions for much of the period from 1956 to 1975. The AMX30B relied on the Obus-G, the family of lighter AFVs fired lightweight HEAT rounds. Even after the CN-105-F1 received a dedicated KE round, the stocks of OCC Mle 62 were large enough that these rounds were still in widespread issue after the Cold War ended.

Sighting and optics – AMX30B

The first design priority behind the concept of the AMX30B was long-range firepower, and its main armament was considered to be one of the best of its type in the world at the time of its adoption. The basic idea behind the AMX30B's fire control system was to give the crew excellent battlefield observation and target acquisition, which permitted rapid engagement and destruction of enemy tanks at long range. The turret's ergonomics allowed the loader to maintain a rapid rate of fire. The commander's rangefinder and the gunner's sights, combined with the inherent accuracy of the CN-105-F1, was expected to give a 75% chance of a first-round hit at 3,000m, or a 90% chance at 2,500m (under daytime battlefield conditions on an unobstructed target). In reality, the tank commander's line of sight on the European battlefield seldom offered more than 2,000m of unobstructed view, at which ranges the AMX30B's fire controls proved very effective.

Infrared devices, which were the most advanced nocturnal combat devices available in the early 1960s, had already been successfully adopted as standard equipment on the AMX13 series in 1964. The night vision equipment in the AMX30B's fire controls was substantially improved over these basic devices, and these were supplied in order to replace the normal daytime optics provided. The potential combat ranges with infrared equipment was much shorter

BELOW A close-up view of the TOP7 (*Tourelleau d'Observation Panoramique No 7* or all-round observation cupola No 7), which gives the tank commander an independent observation capacity. The TOP7 had a hand traverse and the commander could also contra-rotate the turret to line up on to targets he identified (essentially an older form of the hunter-killer doctrine). The TOP7 was a cast structure with a welded roof which incorporated ten M268 direct vision periscopes affording a 360-degree field of view around the tank. It also served as a mounting for the M270 prism head into which the commander's sights fitted. The cupola-mounted 7.62mm ANF1 machine gun was provided with 2,070 rounds, stowed in the turret and the turret stowage bin. The ANF1 was manually operated from inside the cupola by means of an electrical firing switch, and could be remotely cocked or reloaded. The PH9A searchlight seen here is fitted with its infrared filter.

(Fonds Claude Dubarry, Collection du Musée des Blindées de Saumur)

RIGHT AMX30B turret with the night IR vision devices in place. On the right, the detachable PH8B searchlight coupled to the mantlet. The PH8B's IR filter was selected from inside the tank. The OB-23-A IR binocular sight replaced the tank commander's M267 day binocular sight in the M270 prism head located just in front of the commander's hatch on the top of the TOP7 cupola. The M282 gunner's periscopic sight was replaced by the OB-17-A infrared sight. An infrared filter was fitted to the PH9A searchlight on the TOP7's 7.62mm machine-gun mounting. All in all, the preparation for night combat took about 15 minutes for a trained crew. *(Fonds Claude Dubarry, Collection du Musée des Blindées de Saumur)*

RIGHT This is the AMX30B turret set up with day sights. The PH8B is in white light mode, the commander's M267 sight is fitted in the M270 prism head, the slightly smaller M282 gunner's periscopic sight is provided and the PH9A is not equipped with an infrared filter. *(Fonds Claude Dubarry, Collection du Musée des Blindées de Saumur)*

than with 'day' sights but on the AMX30B fitting the infrared sights was a simple process. The day sights were dismounted and carefully stowed, the infrared sights unpacked and installed, and then zeroed. The use of infrared equipment in the AMX30B shared the same limitations seen in other infrared tank fire controls of the day. As one might expect, the use of an infrared projector was easily detected by any vehicle equipped with image intensification or thermal imaging equipment in the 1970s and 1980s.

The range of optical equipment provided for the AMX30B was functional, exemplary for its time – and the careful design of the fire controls and optics gave the AMX30B crew a very good chance of a first-round hit when firing from the halt. The use of a main armament stabilisation system was not seen as a design priority and was never incorporated retroactively into the AMX30B or AMX30B2. The fire on the move capability that a stabilisation system might have offered would certainly have made the AMX30B and AMX30B2 truly elusive targets, but this was never achieved.

The 30-tonne tank was intended to make maximum use of its firepower and to

BELOW The M208 cross turret rangefinder assembly was mounted to the turret roof and was used by the commander. We can see the commander's operating controls here. *(Hugues Acker)*

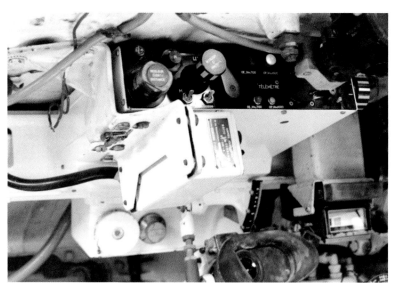

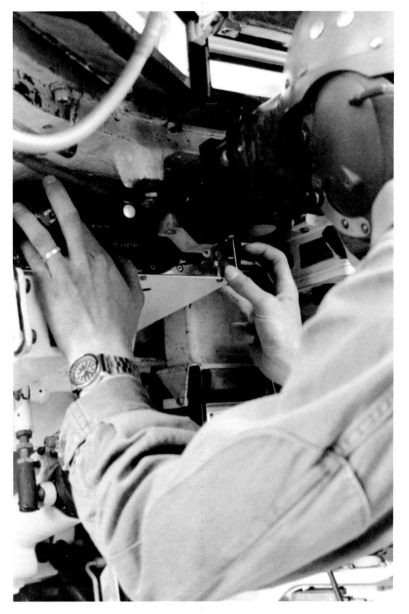

LEFT A tank commander scans for targets with the M208 rangefinder. *(Hugues Acker)*

compensate for its thin armour by conducting engagements at long range, changing position frequently but always firing from the halt. Previous French tank designs like the AMX13 Mle 51 were too small to incorporate anything more than basic gunnery sights. The other side of that developmental coin explored in the decade after 1945 was the AMX50, designed to fire KE rounds to long ranges.

The AMX50 prototypes had proven the feasibility of a turret-mounted rangefinder. Then came the M47, with its complex stereoscopic rangefinder, which required specially selected gunner candidates. The experience of using the M47 pushed DEFA firmly towards the further development of a coincidence rangefinder. The research conducted to provide the AMX50 with a functional rangefinder between 1949 and 1957 was carried forward by APX and SOPELEM to assure the 30-tonne tank a good chance of a first shot hit at long range, and the M208 rangefinder was the fruit of this research.

The M208 employed a 2m base (with 12× magnification and a 3-degree field of vision), and it could be used to identify targets between 600m and 3,500m away under average lighting conditions. It could be used over shorter ranges by the commander as a 6× magnification telescope with a 5-degree field of view to employ the main armament independent of the gunner's sight. The M208's sighting heads were located on either side of the turret just behind the gun mantlet, linked to the gunner's M271 main telescopic gun sight (an articulated monocular telescope mounted in the gun mantlet, on the right-hand side of the main armament). The M271 was normally employed under daylight conditions. It enjoyed a 9-degree field of vision with 8× magnification.

For nocturnal engagements the SOPELEM PH8B 450mm 250-watt white light searchlight/infrared projector was mounted on the left-hand

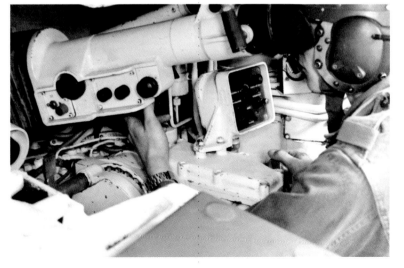

LEFT A gunner using his M271 sight unit in an AMX30B of the *6e Régiment de Dragons* in the late 1970s. *(Hugues Acker)*

side of the gun mantlet, aligned to the 105mm gun. The PH8B had a range of 1,000m in infrared mode and 1,500m in white light mode, and its shutters could be opened and switched between modes from within the turret by means of two filters. The AMX30B's infrared sights depended on this device to illuminate targets. The main gun sight for nocturnal use was the OB-17-A periscopic sight, which incorporated 5.4× magnification, a field of view of 7 degrees and could be elevated -8 to +20 degrees for a maximum effective range of 1,000m.

The AMX30's driver, loader and gunner were all provided with simple observation-type optics, in the form of M223 direct-vision episcopes. Three of these were mounted over the glacis in the front sill of the driver's hatch, providing an overlapping field of view. Naturally, the AMX30B's drivers always preferred driving head out, which required the turret hydraulic traverse to be locked out. In the turret an M223 episcope was located to the right of the gunner's seat in the turret wall permitting a 28 × 95 degree (vertical and horizontal respectively) field of view to the right from the turret, and two M223s were mounted in the turret wall to the left of the loader/radio operator. The loader's pair offered a slightly wider field of vision due to the overlap of fields of vision. These each provided a useful 1× magnification vision capability around the front and sides of the tank when closed down. The loader and gunner each were also equipped with a traversable M282 periscope (with a 26 vertical × 42-degree horizontal field at 1× magnification). The M223 optics were developed by the Clave corporation under the direction of the *Atelier de Puteaux*, and the provision of inclusion of both fixed and traversable vision devices for the gunner was exceptional for the period. The driver's central M223 could be replaced for night driving with the SOPELEM OB-16-A binocular infrared vision device, which permitted the driver to see about 90m in front of his vehicle. The OB-16-A could function as a normal monocular episcope by switching channels, which better suited low-light driving conditions at dawn or at dusk.

The optics and fire controls provided for the 30-tonne tank evolved considerably between the production of the first prototype and the production AMX30B of 1967, in part because

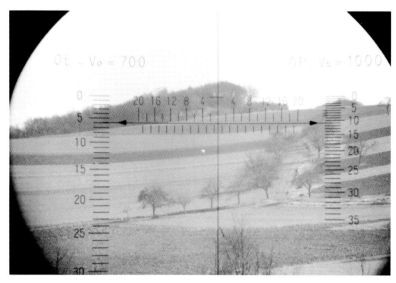

the various components of the turret systems were perfected separately. The commander's TOP7 observation cupola, for example, was not ready before 1966. Due to the height of the TOP7's mounting on the turret roof a field of view of 45 × 125 degrees was possible with each periscope along with an inclination of -6 to +6 degrees. To the immediate front right of the TOP7 on the first 160 turrets manufactured was a small aperture in the turret roof for an M259 azimuth indication device, which was deleted during production as it proved unnecessary.

The roof of the TOP7 incorporated the commander's hatch, a fixed M270 armoured periscope mounting for using the primary and secondary armament and the PH9A cupola white light/infrared searchlight. The M270 prism served as the external optical device reflecting into the M267 commander's binocular

ABOVE The M271 sight graticule. Note the markings for the OE (high-explosive) on the left and OP (*obus perforant* – armour piercing – in this case the OCC-F1 alias Obus-G) on the right with velocity data. The top of the graticule betrays the original DEFA designation for the CN-105-F1 – '105mm D1512'.
(Hugues Acker)

LEFT The TOP7 was employed in its original form right up to 1997 when the last AMX30B regiment (the *2e Régiment de Dragons*) turned in their mounts. The M270 prism seen here was used with the M267 binocular day sight. This photo, showing the cupola without the commander's AN F1 7.62mm machine gun fitted, was taken very close to the type's last days in service.
(Zurich 2RD)

periscope and collected light for the OB-23-A cupola infrared sight. Both of these devices shared the M270 prism for aiming the AN F1 cupola machine gun and the coaxial 12.7mm or 20mm armament (the former through mechanical linkage, the latter through an electric servo linkage). The M270 mounting traversed with the cupola roof through 360 degrees and incorporated an elevation of -10 degrees to +45 degrees. The M270 prism had a 100 Mil field of

vision and could be elevated with a handwheel mounted inside the cupola (which also served to elevate the cupola machine gun).

The M267 commander's binocular periscope incorporated 10× magnification enabling the TOP7 to be used as a target-acquisition device at long range independent of the main turret gun sight and optical rangefinder unit. The OB-23-A cupola infrared sight was used in conjunction with the cupola SOPELEM PH9A searchlight in infrared mode. It was used to aim the AMX30B's secondary and cupola armament and as a general nocturnal observation device for the tank commander. As an aside, the commander might also have recourse to the use of handheld OB-24-A 4× magnification infrared binoculars, which could be used for infrared surveillance (and which could detect

BELOW The T105M turret developed for the AMX30B2. Since this drawing lacks the armoured mounting for the AN F1 7.62mm machine gun, it was very likely derived from the prototype. The T105M layout saw the wholesale replacement of the M208 rangefinder and M271 sight among others, replaced with electronically controlled optics. The secondary optics were retained in the spirit of economy. (Collection Thomas Seignon)

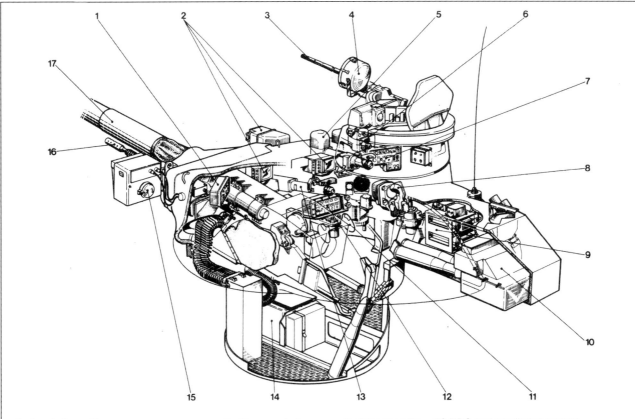

1 Loader's position	**10** NBC and ventilation system
2 DIVT13 low-light television camera computer and monitors (gunner and commander)	**11** Gunner's traverse and firing control
	12 Gunner's position
3 ANF1 7.62mm machine gun and armoured cupola mounting	**13** M581 gunner's primary sight
	14 Turret main electrical junction box
4 PH9A searchlight	**15** PH8B mantlet white light/IR searchlight
5 M496 secondary sight	**16** CN-20-F2 20mm coaxial weapon
6 Commander's monocular telescope for M496 sight	**17** CN-105-F1 105mm main armament
7 TJN223 telescope (for commander's OB49 sight)	
8 Commander's traverse and gunnery hand control	
9 TR-VP-13-A and TR-VP-113-A radio installation	

infrared projection) in a 15-degree field of vision at ranges of 300–1,000m. The AMX30B crew enjoyed a functional range of night vision devices, but all infrared night vision equipment of the 1950s and 1960s suffered from a series of limitations when compared to later image intensification or thermal vision systems – and in time proved readily detectable by the latter.

In summary, the AMX30B had an excellent suite of optics for the late 1960s – certainly among the best in NATO. The TOP7 was of course vulnerable to a host of weapons, but it allowed the tank commander an unparalleled view of the battlefield (it is still lauded as the best observation cupola ever in the French Army). The CN-105-F1 gun allied to the M208/M271 proved to be a reliable and extremely accurate combination. A well-trained AMX30B crew could find and engage targets at long range, could maintain a high rate of fire and enjoyed simple and effective sights and observation equipment.

The AMX30B2's COTAC fire control system

The AMX30B2's COTAC (*COnduite de Tir Automatisée pour Char*) electronic fire control system was adapted from the system employed in the AMX32's first pattern turret. It was designed to give high first-round hit potential when firing the 105mm gun at maximum range at fixed or at moving targets. It was introduced in 1982 at the same time as a high-pressure KE armour-piercing round for the CN-105-F1. The original COTAC system

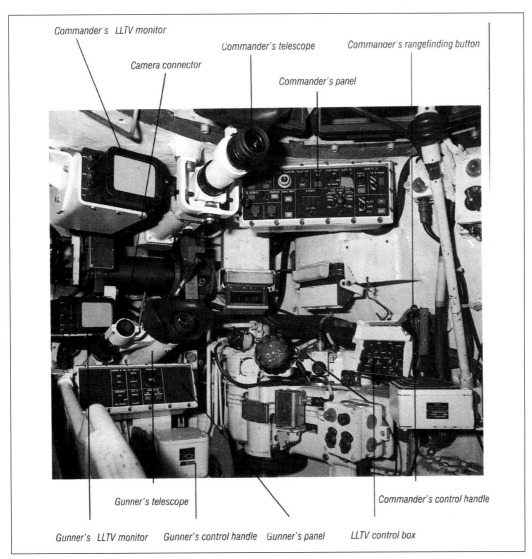

Commander's LLTV monitor
Camera connector
Commander's telescope
Commander's rangefinding button
Commander's panel
Gunner's telescope
Gunner's LLTV monitor
Gunner's control handle
Gunner's panel
LLTV control box
Commander's control handle

LEFT The principal gunner's and commander's controls employed with the COTAC system in the AMX30B2. The commander could override the gunner's controls during an engagement if necessary. *(Nexter)*

RIGHT The commander's position of an AMX30B2 employed by the FORAD detachment at Mailly around 2007. On the left, we can see the monitor for the DIVT16 thermal camera, with the commander's telescope (for M581 and the M496 sights). To its right, the commander's panel to input for weather conditions, altitude and ambient temperature. These were entered daily to give the COTAC system the best possible chance of a first-round hit. *(Zurich 2RD)*

included an image intensifier system which permitted its use by day or by night. The low-light DIVT13 camera was mounted coaxially to the main armament on the right side of the gun mantlet, and was replaced on late AMX30B2s with the DIVT16 CASTOR thermal camera (and on the 1996 AMX30B2 Brennus conversions with the DIVT18).

The COTAC system was designed to calculate all gunnery parameters in order to accurately sight and destroy a target within a minimal engagement time. It accounted for atmospheric conditions through a daily crew data input, as well as real-time wind speed and direction, altitude, temperature and ambient humidity. The gunner's own tank's position and trunnion angle, the ammunition ballistics and the gun's characteristics adjusted the aiming graticule in the gunner's sight automatically.

The COTAC system incorporated the M581 sight with its associated optics, the M421

LEFT The gunner's position in the same tank. We can see the M581 telescopic sight, which was designed to fit into the same mounting as the older M271 used on the AMX30B. The red-handled gunner's control was employed to elevate or depress the main armament as well as to control turret traverse. *(Zurich 2RD)*

compensator, the M550 laser rangefinder, the M579 fire control and filter unit (which included a gyroscope, firing matrices and trunnion tilt sensors), the primary junction unit, the gunner's and commander's control panels, the gunner's firing control and the image intensification camera. The laser was accurate to within 5m at all ranges between 300 and 9,600m. An engagement sequence might be conducted as follows, taking mere seconds between target acquisition and the first shot:

The target would be identified by the commander through the TOP7 cupola's OB49/TJN223 sight, which was then acquired by the gunner in his M581 sight. The commander would order the gunner to range the target, and the M550 rangefinder integral to the M581 unit would establish the range. A stationary target would be ranged for approximately 1 second using the measure button on the gunner's hand control. A mobile target would be measured for up to 3 seconds. As soon as the gunner released the measure control, the range in metres would appear on the commander's panel. A flashing range figure would indicate that the laser had bounced off an object between the rangefinder and the target, such as vegetation. The tank commander would in such an instance make a judgement if the distance (known as the first echo) could be used for the first shot, or if the sequence would be simply repeated. Once the distance was confirmed to the gunner verbally by the commander, the gunner moved his aiming reticule in the gun sight on to the target, awaiting the order to fire.

The commander could also employ the laser rangefinder himself, and he could override the whole fire control sequence in an emergency. The M581 sight could be used in full COTAC mode, in a simple mode with only the rangefinder function activated or as a simple gun sight should the COTAC system be rendered inoperable. The COTAC system proved simple, reliable and effective in use and represented a major advance over the optical rangefinder-based fire controls of the AMX30B. Combined with an experienced loader, the COTAC system offered a high rate of fire as well as a high hit probability. The gunner's and commander's positions were each provided with a monitor for observation with the DIVT13,

ABOVE The AMX30B2 retained the TOP7 cupola, re-equipped with the OB49 day/night sight and a new armoured mounting for the AN F1 7.62mm machine gun. *(Zurich 2RD)*

which was also used in conjunction with the commander's and gunner's gunnery sights. In the last batches of AMX30B2s converted, and in rebuilt early AMX30B2s, the more effective DIVT16 CASTOR (developed by Thomson-CSF and subsequently manufactured by Thales) thermal camera replaced the DIVT13. The DIVT16 proved to be an excellent and long-lived piece of equipment which served the post-Cold War AMX30B2 park for three decades.

BELOW The AMX30B2's loader's position was also altered considerably, with the reduction of ready ammunition to a single round, and the radio equipment moved to the right side of the turret bustle. *(Zurich 2RD)*

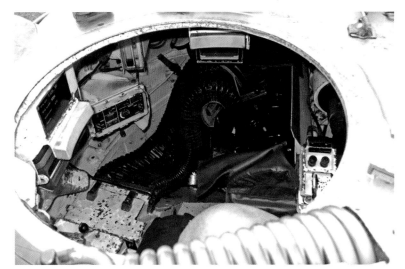

Chapter Four

Maintaining the AMX30

One of the greatest strengths of the Cold War French Army was the integration of the armaments industry with the army's network of maintenance facilities. The AMX30B enjoyed a rigorous maintenance schedule observed at the crew and unit level. Every sub-system in the AMX30B and in its successors and variants were subject to detailed and well observed maintenance procedures to allow maximum availability.

OPPOSITE The full servicing of an AMX30B's turret was performed on an as-needed basis. With the turret removed, the hull interior was meticulously cleaned. The opportunity was taken to carefully inspect the rotary union supplying power to the turret's electrical system. *(Hugues Acker)*

Servicing and non-technical inspections

Vehicle commander

In the French Army the tank commander carried the ultimate responsibility for his machine and for the lives of his crew – and a well-maintained tank offered the best chance for survival on the battlefield. The tank commander oversaw the work of his subordinates and served as an extra body permitting the rapid completion of the entire daily task of vehicle maintenance (by relieving the driver or other crewmen).

Commander's maintenance tasks:

- Fluid levels (with the driver).
- Turret traverse function and gun elevation mechanisms (with the gunner).
- Commander's traverse override control.
- Gun travel lock and breech safety lock (with the loader).
- Recoil mechanism for the main armament, verifying recoil travel (lubricate the recoil cylinder).
- Function of the case extractor during recoil travel.
- Hand traverse and contra-rotation function of the TOP7 cupola.
- Inspection and test of the NBC system prior to each sortie or every month. This required that he set the switches to correct mode, testing that the fans were functioning correctly. He then ensured the liquid level in the manometer was right, then tested the overpressure and filtration functions.

As one might expect, the commander tested out the wireless set prior to any sortie and monthly. A good commander was quite familiar with

BELOW The *DuGuesclin* of the *6e Régiment de Dragons* in the course of having its complete turret hoisted off the hull in regimental workshops. The tightly packed turret basket and inflatable turret ring seal are both visible. The hoist is probably a 30-tonne unit. Turret mechanics, known as *tourellistes* in the French Army, could thoroughly inspect the dismounted turret, its weapons and most of its systems on a stand like this one. *(Hugues Acker)*

any maintenance task, but might be burdened with orders groups, administrative functions or other tasks by the platoon commander. Platoon or squadron commanders had a substantial communication network to look over, as well as map reading and replenishments to coordinate.

Gunner

The gunner's role in maintaining the AMX30B was focused primarily on the CN-105-F1, the secondary armament (12.7mm or 20mm CN-20-F2) and the associated sighting systems. Leaving the sighting and fire controls aside, the gunner was tasked with a detailed inspection of the main armament's condition prior to any firing period, at the conclusion of any firing period and otherwise weekly in barracks. This extended to the turret's hydraulic systems, which he would check at the turret traverse control box, the cupola contra-rotation control box and along all the lines. The gunner was also in charge of ensuring that the cupola optics wiper reservoir was filled with windscreen washer fluid. The gunner checked the oil levels

in the hydraulic traverse system, the recoil cylinder and in the air compressor.

The gunner was also responsible for the maintenance of a properly stowed and secured turret bustle bin. Inside the turret, the gunner performed a daily inspection of the fighting compartment fire extinguishing system and regularly tested its alarm. Before firing and weekly, he was responsible for testing the gun's mechanical elevation handwheel, feeling carefully for any stops or tight points, which might have indicated a damaged gear or bearing. He was also to test the hydraulic traverse and elevation controls at low speed, including the full elevation and depression of the gun. The CN-105-F1 gun itself required the gunner's inspection prior to firing and weekly for general condition, a careful going-over to ensure no components were loose (repeated at the end of each range period as well), an inspection of the electrical firing circuit and its switch, the presence and condition of all the gun's accessories and finally the inspection of the recoil system for any leaks and the system's oil level.

BELOW The gunner's position of an AMX30B preserved in the Musée des Blindées de Saumur. The red gunner's traverse and elevation handle is located directly below the M271 telescopic sight. *(Thomas Seignon)*

The loader–wireless operator

As on any battle tank, the loader (*Radio Chargeur*, 'RC' or *Romeo Charlie*) was tasked with a mixture of duties, spread between the care of the ammunition and the wireless systems – often with further jobs associated with cooking the crew's meals or tasks delegated by the vehicle commander. He was a third set of eyes watching for any signs of oil leakages in the turret hydraulics and he shared responsibility with the driver for the inspection of the air cleaners and the regular topping-up of their oil levels. Air cleaner inspection could be made very frequently if the AMX30B was operating in dusty conditions. He was responsible for the inspection of the gun's thermal sleeve and its muzzle cover, as well as for all of the exterior stowage. He was expected to check the condition of the case ejection port, the fume evacuation port and the wireless antennae bases. He was required to regularly grease the case ejection port hinge, the commander's hatch and loader's hatch hinges and to test the hull escape hatch and driver's hatch function with the driver. He shared responsibility for inspecting the track tension and the idler wheel adjustment with the driver, for checking the tightness of every bolt securing the road wheels, sprockets, idlers and return rollers before departures and at every halt. The track pins would be inspected for any missing retention circlips. Each wheel hub grease nipple would be inspected for damage and tightness.

He would check over the turret bustle and hull ammunition stowage prior to any departure and at the end of any sortie.

The loader was also charged with tasks associated with the main armament. He was responsible for ensuring that the gun was unloaded at the end of each firing period, and (with the commander present) to ensure that the case extractor functioned properly through the recoil mechanism's travel. The compressed air system, its gauges and its tank – absolutely essential to the CN-105-F1's fume evacuation system and vital for inflating the pneumatic fording seals – was checked rigorously every week by the loader, as well as before and after any sortie or range period. Its associated compressed air tank was purged weekly, prior to firing and after each range period. The 105mm ammunition was carefully kept dry by regular wiping during loading and unloading by the loader. The loader also made a regular inspection of all ammunition racks and each round's secure placement after loading. Ammunition was normally moved from the front hull rack into the turret bustle rack as the ammunition load was fired, a task assured by carefully traversing the turret and transferring the stowed rounds.

The loader-operator was tasked with the basic maintenance of the radio system. The AM84 amplifier unit was carefully inspected for the effects of vibration loosening it from its mounting. Naturally, because the wireless equipment was sensitive to humidity, any time the tank was washed down with high-pressure hoses, the inflatable seals were all inflated. Wireless contactors and fuses were also checked monthly, as were the turret's wireless mountings, insulators and shock-absorbent mountings (which had to be checked for tightness). The antenna leads, connecting wires and fuse boxes were also inspected monthly. The antenna bases, sections and isolators (made up of lower and upper halves) were all inspected prior to any sortie. It was important to make sure that no grease or paint got on to the isolators. Each of the four control boxes and three relay boxes associated with the wireless system were checked monthly. The wireless/interphone headsets were inspected prior to use by each crew member. Particular care

BELOW The loader was responsible for maintaining the ammunition racks. This photo shows drill rounds stowed in the turret bustle. While the picture shows an AMX30B2, the ammunition stowage in the front right of the hull and in the left side of the turret bustle did not change between the AMX30B and AMX30B2. *(Zurich 2RD)*

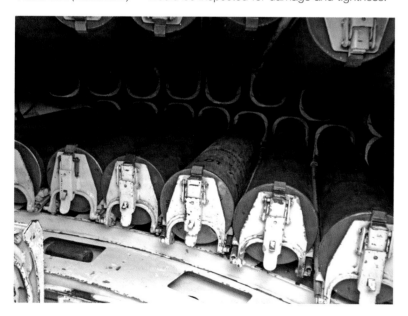

was paid to examining the wire leads for any tangles or damage. Each relay and control box was inspected and the potentiometers and commutators checked.

Shared maintenance tasks

Turret electronics and fire controls

In the AMX30B the gunner, loader and commander were responsible for checking over each of the turret control panels. Maintenance of the turret optics was a simple process followed regularly by the commander and the gunner. The M282 rangefinder, gunner's sight, commander's binocular sight, infrared sight (if fitted) and PH8B mantlet infrared/white light projector were all zeroed daily at the start of operations. The frequently removed optics (which included the infrared sight, the commander's M267 day binocular sight, the TOP7's M282 episcopes, and the M223 episcopes mounted in the turret walls and driver's compartment) could be removed and

wiped after submerged crossings. The rest of the AMX30B's turret optics were expected to be cleaned and maintained in place without disassembly by the crew. Each sight reticule, elevation indicator and lighting system would be checked for proper function prior to a range period by the gunner or the commander. Any defect could be rectified at the regimental workshop level by specialised technicians.

Weapons maintenance

The commander's 7.62mm AN F1 machine gun was checked over and zeroed by the commander, its remote cocking mechanism and electric firing switches were tested, and its semicircular feed tray was inspected for any damage and loaded with ammunition belts. As with any weapon of this type, regular cleaning of the barrel and receiver, lubrication and inspection of the gas system was performed daily. The weapon was easily dismounted for inspection and cleaning, and was well regarded enough to be retained in an improved mounting on the AMX30B2.

The CN-105-F1 105mm gun barrel was

ABOVE Cleaning out the bore was a procedure followed after every range period, and one that could include the whole crew. The bore-cleaning rod was screwed together from sections carried on the turret roof between the loader's hatch and the commander's cupola. *(Fonds Claude Dubarry, Collection du Musée des Blindées de Saumur)*

ABOVE In this instance it appears that two full crews are on hand, counting the photographer. *(Thomas Seignon)*

inspected weekly by the commander and gunner for any signs of corrosion or any traces of wear. The bore was swabbed clean after every range period so that no traces of powder residue were present, while the rifling was inspected prior to the muzzle cover being fitted. This operation often involved the whole crew, but could be performed by the gunner and loader if necessary. The breech mechanism was lightly oiled and the firing needle was inspected. The breech block and firing circuit were inspected monthly. The fume evacuator system was tested prior to any range period and monthly.

The early AMX30B's coaxial 12.7mm machine gun (referred to in some official publications as the *Mitrailleuse de Cal .50*) was a licence-produced M2 Browning. Its maintenance regimen fell to the gunner and loader, but the weapon was exceptionally robust and easily cared for. Prior to any range period and weekly, the gun mount was inspected, the weapon was removed, stripped, inspected, lubricated, reassembled and zeroed. Its ammunition feed box and belts were carefully inspected and loaded. Finally, its tool kit was checked over and carefully packed away. The turret smoke dischargers were unloaded, dismounted, stripped and lubricated,

reassembled and remounted before finally being loaded prior to use and monthly. When the CN-20-F2 20mm cannon secondary armament was introduced the inspection regimen was similar, but special care had to be paid to the handling of the 20mm belts and checking of the ammunition feed tracks (which were easily bent, causing stoppages).

Driver

While the entire crew participated in the AMX30B's maintenance, by any measure the heaviest of these tasks fell upon the driver – a daily regimen much repeated which might take 2 hours unassisted. Driving the AMX30B took skill and practice, which could only be maintained in a relatively low number of allowed operating hours each month. As a result, in a conscript army, driver training had to be carefully followed to avoid damaging the AMX30B's Gravina centrifugal clutch and its BV-5-SD-200D transmission in general. When the AMX30B was introduced in 1967, a call went out for experienced AMX13 drivers (used to an even tighter matrixed manual gear shift) to train on the new tank and develop best practices. In the hands of an experienced driver, and with a well-followed maintenance routine, gearbox issues were minimised.

The largest and most spectacular powertrain-related tasks involved the engine, gearbox or more commonly the complete powertrain being hoisted out of the engine compartment following the careful removal of the rear hull plate and engine decks. This was a task performed by the regimental workshops section or in the field under exceptional conditions by the support company's AMX30D. Nonetheless the vehicle driver and other members of the crew were expected to assist this kind of task under the direction of the mechanics. Such events were actually quite infrequent, and maintenance normally consisted of far more mundane verifications and routine checks.

The AMX30B driver's substantial burden of maintenance tasks was precisely set out in numerous French Army manuals, and would have been similar for most of its derivatives. He was to inspect for water, coolant or oil leaks beneath the hull at all stops, followed by a quick inspection of the shock absorbers and suspension units for oil leaks. He was expected to open the rear access hatch to check the engine compartment prior to departure (and at day's end) to inspect for any leaks around the oil pump. At the same time, he would inspect the final drives' brake lines and greasing points for any traces of leaking oil.

The driver was expected to check for coolant leaks in the radiator system (inspected via the upper access hatch), for any fuel leaks from the front right diesel fuel tank and any oil leaks in the brakes and steering controls. He would then check the fuel level, coolant level (with possible addition of antifreeze). The engine oil level was checked in the presence of the tank commander. The gearbox and final drive oil levels were then verified, the engine oil bath air filters were inspected with the loader-operator and the steering and brake system oil levels were inspected. Finally, the driver inspected the battery electrolyte level, ensuring at the same time that the connectors were in good condition and free of any collected debris.

More mundane tasks had to be followed prior to any departure from the tank park. The driver had to check periscope washer fluid level, account for all pioneer tools and exterior hull stowage and check the hull stowage bins and their locks. He would check the track guards, vehicle lights and exhaust system for any damage inflicted by cross-country operation. He would check the fasteners on all inspection plates and – with the loader – the function of the driver's hatch and hull escape hatch. He would check and grease (as required) the fuel filler covers, the driver's hatch mechanism and engine deck hatch hinges.

At every stop the driver would check the suspension with the loader as meticulously as possible. He would inspect the torsion bars and suspension arms, ensuring that none were poking proud from their mounts in the sides of the hull. Track tension was a particularly vital condition on any tank, and it was checked constantly to avoid the dreaded problem of shedding a track. The sprocket teeth and final drives were also checked nearly as often. Finally the driver would check by feel for any heating of the wheel hubs and return roller hubs – a sure

ABOVE Photographed in the workshops of the *6e Régiment de Dragons* in the late 1970s, a complete powerpack change is undertaken by a maintenance crew. The engine was completely uncoupled from its mountings, the final drives, fuel supply and electrical supply. The tank's driver and other crew members might be expected to help with this task, which was undertaken on an as-needed basis, or otherwise upon reaching a prescribed number of operating hours. *(Hugues Acker)*

BELOW The engine and transmission was suspended on a three-point jig which permitted a balanced lift with the whole powerpack. A similar jig was employed with the AMX30D's onboard lifting crane. *(Hugues Acker)*

sign of worn bearings. Once a month (which equated to 25 hours of use under normal conditions) or after any operation in muddy conditions, the suspension required a complete greasing. This included the shock-absorber joints, swing arms, idler tensioning assembly, track return rollers, road wheel bearings and the sprocket bearings.

Inside his own crew position, the driver's controls were tested prior to each operation of the vehicle, or weekly. This included all dashboard panel control functions, the function of the interior lights, running lights and siren. He would check the proper function of his steering levers and gear-change lever, his accelerator pedal and accelerator hand control. The overpressure valve on the right front fuel tank and the crew compartment heater were tested (vital if the tank was not to turn into a rolling refrigerator in winter exercises). The driver would check the seating of the episcopes and wipe each with a clean, dry cloth. He would dismount and carefully dry any of the M223 and M282 episcopes and dry thoroughly if any water had penetrated between the armour plate and the episcope body in inclement conditions. He would then clean the body and grease the seal seat of each prior to reinstallation. In complete darkness, which might mean in the earliest

ABOVE The underside of the HS-110 engine. The HS-110 was a reliable flat 12-cylinder diesel engine, referred to in contemporary literature as a multi-fuel engine, but never run in practice on any fuel other than diesel. *(Hugues Acker)*

RIGHT The empty engine compartment of an AMX30B. We can see the two inspection holes in the hull floor, the panel for turret aspiration, the upright fuel tanks and the large final drives. *(STAT)*

hours or within the confines of his compartment with the vehicle interior lights switched off, he would test the function of the infrared driving periscope. He would finally test the episcope washers before any operation of the vehicle.

Every month the driver would grease the adjustable seat joints, the steering levers, the gear-change lever, the accelerator pedal and lever and the brake pedal and lever. He also had a schedule of weekly tasks to be followed more frequently if the vehicle was being employed on operations. The battery compartments were opened up and the eight 12V batteries were checked over, the battery connections were inspected for proper seating, the plugs for any corrosion or obstruction from debris. Any defects were reported to the vehicle commander. The radiator was checked by lifting the engine cover, any debris in the radiator grilles was carefully removed and the fans and their drives were closely examined. Naturally, the task of helping the driver often fell to the loader, but any defects were reported to the squadron headquarters if the maintenance section was needed. The availability of one of the regiment's AMX30Ds to speed the heaviest tasks was always welcomed – but the onus was always on the driver to know his tank's mechanical condition.

ABOVE The exterior of the AMX30B's driving compartment. The three driving periscopes can be seen; the sheet metal combing was used to anchor a foul weather hood made out of clear plastic sheet. It was seldom used. *(Zurich 2RD)*

BELOW LEFT The interior of the AMX30B's driver's compartment photographed on a preserved vehicle at Saumur. The steering levers are visible, as are the accelerator and clutch pedals and the gear shift, situated between the driver's legs. *(Thomas Seignon)*

BELOW RIGHT This arrangement changed considerably in the AMX30B2 – a view through the loader's position is seen here. We can see the steering wheel that replaced the steering levers, rendering it much easier to drive. *(Zurich 2RD)*

Chapter Five

The AMX30 family of vehicles

DEFA planned from the outset of the 30-ton tank programme to employ it as the basis for a family of variants. This embraced a lineage of recovery, artillery and engineer's vehicles manufactured on the AMX30 chassis that represented a massive investment in the army's capabilities. Among them were the stalwart AMX30D, the imposing AMX30P Pluton missile launcher, the advanced AMX30R Roland and the 155mm self-propelled AU F1.

OPPOSITE The AMX30 family of vehicles: an EBG photographed in 2014 prior to the Bastille Day parade. (Jerome Hadacek)

ABOVE Registered as W000-870, the AMX30D prototype is seen here during trials at the end of the 1960s. The general layout of this well-designed vehicle was already nearly finalised. The AMX30D was highly esteemed by its users and enjoyed a career spanning nearly 50 years. *(Fonds Claude Dubarry, Collection du Musée des Blindées de Saumur)*

BELOW The hydraulically controlled Griffet crane was a key asset for powerplant replacement operations. The crane could traverse under load 240 degrees (from the 10 o'clock to the 6 o'clock position). A maximum of 15 tonnes could be lifted directly ahead of the vehicle and the crane could traverse with a suspended load of up to 12 tonnes. *(Fonds Claude Dubarry, Collection du Musée des Blindées de Saumur)*

The full picture of all planned AMX30 variants considered by DEFA, DTAT and GIAT will probably not be available to the researcher for years to come. The variants that were approved for production included specialised recovery, artillery and engineers' vehicles for use in armoured formations equipped with the AMX30B or AMX30B2 battle tanks. These specialised programmes were supplemented after the Cold War with conversion programmes to upgrade existing capabilities or to make use of surplus hulls in the spirit of economy. Many of the conversion programmes were shared between ARE and the army's workshops at Gien to minimise cost.

AMX30D recovery vehicle

When the AMX30B entered service in early 1967 it was supported by the Sherman-based M74 tank recovery vehicle supplied through the MAP programme to support the M47 Patton. This vehicle remained in French service until 1976, being replaced by a specially designed recovery AMX30D (D for *Depannage* or recovery) variant based on the AMX30 chassis. The requirements set out for the recovery variant included the provision of a central winch and a powerful crane to permit powerpack changes in the field. It had to be capable of submerged crossing and deep-

water fording just like the AMX30B. It also had to include an NBC system and provisions to carry a complete powerpack on the rear engine decking. The AMX30D prototype was built around a central winch compartment. It was equipped with a hydraulic crane and an earth anchor-cum-dozer blade. The prototype was built at ARE and received the registration W000-870, undergoing very successful trials in 1970.

The entire AMX30D hull was laid out around a powerful main winch drum, which could execute recoveries from the front of the machine. It was easily capable of recovering vehicles weighing up to 45 tonnes, and with pulleys or with two AMX30Ds working in tandem could recover heavier vehicles. Paired with an integral dozer/earth anchor and its hydraulic crane, the AMX30D's winch made it a true workhorse. The rear engine deck was equipped as a pannier to carry a full welding kit or a complete AMX30 series powerpack. The AMX30D was crewed by four men: a driver, vehicle commander and two mechanics. The commander was provided with a TOP7 cupola

ABOVE The AMX30D could carry a complete AMX30B powerpack on the rear deck. *(Fonds Claude Dubarry, Collection du Musée des Blindées de Saumur)*

LEFT The AMX30D could also carry a full oxyacetylene welding set. *(Fonds Claude Dubarry, Collection du Musée des Blindées de Saumur)*

RIGHT The AMX30D could ford to the same depths as the AMX30B battle tanks it supported. The AMX30D was a vital asset during a submerged crossing with its ability to winch a broken-down tank off a river bottom. Crossings were supported by combat divers, tasked with surveying the river bed and attaching rescue lines if necessary. *(Fonds Claude Dubarry, Collection du Musée des Blindées de Saumur)*

RIGHT This AMX30D is carrying a complete powerpack on its rear deck. It had been removed using the crane and a dedicated triangular lifting jig, which can also be seen above the powerpack. The front auxiliary winch is clearly visible above the blade, which is elevated in travelling position. *(Fonds Claude Dubarry, Collection du Musée des Blindées de Saumur)*

RIGHT The AMX30D's front winch had a maximum line pull of 4,500kg. The winch cable was 120m long with a diameter of 11.2mm. *(Fonds Claude Dubarry, Collection du Musée des Blindées de Saumur)*

ABOVE This AMX30D displays its complete set of onboard recovery and repair equipment (hooks, pulley-blocks, slings, towing triangle, lifting devices and so on). The engine deck pannier is empty, and both the AN F1 7.62mm cupola machine gun and the PH9A searchlight have been dismounted from the TOP7 cupola. *(Fonds Claude Dubarry, Collection du Musée des Blindées de Saumur)*

ABOVE RIGHT Three AMX30Ds were deployed in support of the *6e Division Légère Blindée* in Operation Daguet in 1991. *(Thomas Seignon)*

RIGHT One of three AMX30Ds deployed in support of the 4e RD during the 1991 Gulf War seen back in transit in France on a railway flatcar in July 1991. Upon its return it was freshly repainted and readied for the Bastille Day parade in Paris down the famous Champs-Élysées. *(Fonds Claude Dubarry, Collection du Musée des Blindées de Saumur)*

LEFT The ex-German, Second World War vintage flatcar (the SSyms type once developed by the Reichsbahn for Panther tanks) seen here is an upgraded version of a 1943 product. In 1991 a number of these flatcars were still in use by the army railway service. *(Fonds Claude Dubarry, Collection du Musée des Blindées de Saumur)*

ABOVE An AMX30D in the last part of its active service life as a FORAD unit maintenance vehicle at the Mailly training centre. The light grey and black striped camo scheme was typical for original FORAD fleet vehicles.
(Jerome Hadacek)

RIGHT The AMX30P was crewed by an officer (*chef de pièce*) who commanded the artillery system and three non-commissioned officers. These included the driver (normally a corporal), the *grutier-pointeur* (crane operator-missile aimer) and the *operateur-controlleur* (missile operator-controller). The driver sat in a central position in the hull front with the *chef de pièce* immediately behind him, equipped with a simple-vision cupola.
(Fonds Claude Dubarry, Collection du Musée des Blindées de Saumur)

derived from the type mounted on the AMX30B. Its relatively light weight and versatility made it an ideal vehicle for issue in all armoured, artillery and engineers' regiments that employed derivatives of the AMX30 family. The basic design of the AMX30D was so successful that it has remained in use up to the present day and will be retained as long as other AMX30 family vehicles remain in service.

AMX30 Pluton

One of the most important objectives in French defence policy after the end of the Second World War was the development of an independent nuclear arsenal. The army's nuclear capability prior to 1966 was assured by US-supplied Honest John tactical nuclear missiles, weapons which were effectively denied their warheads after 1966. The replacement for the Honest John was the Pluton, France's first and thus far only tactical nuclear missile adopted for service. The AMX30 Pluton (AMX30P) tactical nuclear missile erector-launcher was derived directly from the hull

RIGHT The centreline of the hull featured the hydraulically raised missile carrier/firing unit carried over the engine decks. An onboard goniometer was provided to calibrate the firing ramp in elevation and to calibrate the inertial guidance system. This was located in the left-hand side of the hull in a welded plate housing immediately in front of the *grutier-pointeur* (who sat on the left side of the crew compartment behind the commander). The *operateur-controlleur* sat to the right of the *grutier-pointeur*, both sharing the simple hinged hatch immediately above the *grutier-pointeur*'s seat to enter or exit the vehicle. *(Fonds Claude Dubarry, Collection du Musée des Blindées de Saumur)*

design adopted for the AMX30D. The Pluton launcher also incorporated an onboard crane unit pivoting on the front right corner of the hull. A hydraulic ram was included on the centreline extending over the engine decks to raise the firing unit, which was mounted on a triangular ramp pivoting at the hull rear corners. The hull design incorporated an auxiliary power unit (GAP – *groupe auxiliaire de puissance*) to power the vehicle's hydraulics and electronics. Loaded with its missile, the AMX30P weighed 38 tonnes. Initial tests were conducted with a prototype hull including cross-country trials at the new base area at Canjuers, and at Bourges. Mishaps involving severe damage to an inert missile proved the limitations of carrying the missile loaded at speed across country. One

LEFT The AMX30P changed very little over 19 years in service and each vehicle received meticulous care and maintenance. The AMX30P served not only as the artillery's most powerful weapon, but also as a widely publicised symbol of France's independent nuclear might. The layout of the vehicle was simple and was designed to permit the crew to load, aim and fire the missile in any terrain. Mobility of the AMX30P when unloaded was very similar to the AMX30D. *(Fonds Claude Dubarry, Collection du Musée des Blindées de Saumur)*

trial which tested the missile launcher elevated to 80cm while crossing rough ground (to determine the effect of vibration) saw an inert test missile snap in half, which caused the redesign of the launch container and design changes to the AMX30P launch vehicle. Two prototypes of the AMX30P were produced in 1972 and the 40 production vehicles were ordered to a slightly improved pattern in the following year.

The Pluton missile itself was designed under the government's direction by the Nord Aviation and Sud Aviation joint company *Société National Industrielle Aérospatiale* (SNIAS – the nascent form of today's *Aérospatiale*, which absorbed both companies) from 1968, with the expectation of replacing the Honest John tactical nuclear missile with a French-designed missile and warhead. SNIAS collectively could draw on a huge amount of experience in guided and unguided missile design, which stretched back to German research captured in 1945 and built upon by French designers with great success throughout the ensuing two decades. Construction and evaluation of missile prototypes proceeded quickly, and the first test launch of the Pluton was made in 1970. The production missile could be fired to a minimum range of 17km and to a maximum range of 120km, which restricted its delivery to targets on French or West German soil. In battle the missile would have been employed to hit enemy assembly areas or to destroy enemy mechanised columns. The Pluton missile's accurate deployment was supported with unmanned reconnaissance drones, battlefield radar and by the provision of the most modern fire controls available.

The AMX30P launch vehicle was operated in batteries of two erector-launch vehicles. Each nuclear artillery battery included a host of support personnel and vehicles. Each Pluton missile was carried in two loads in a pair of GBC8K long-wheelbase trucks for normal transportation. The erector-launcher would be

BELOW An unladen AMX30P moves up to rendezvous with its missile trucks. When loaded the vehicle was top heavy and was not expected to move with the launcher elevated apart from in exceptional circumstances. The hull was thinly armoured to resist shell splinters, much of the vehicle's weight being taken up by the missile launching system's solid-state electronics, the extensive hydraulic system and equipment associated with the missile fire controls. *(Fonds Claude Dubarry, Collection du Musée des Blindées de Saumur)*

LEFT Girding for Armageddon. The Pluton missile's launcher unit, including the unarmed rocket and its firing container, was loaded first. The entire operation was accomplished with the onboard crane. (Fonds Claude Dubarry, Collection du Musée des Blindées de Saumur)

parked in a lay-up position to load its missile, and the missile-carrying trucks would park alongside. The AMX30P was designed to load and assemble the Pluton in 45 minutes using its onboard crane. The large firing unit, which incorporated the missile body, was loaded first with the onboard crane on to the erector platform. The warhead was loaded second and fixed to the missile in situ, using the crane.

The plutonium core was normally added to the warhead on the carrier lorry prior to its installation by the AMX30P crew. Within 10 minutes of the loading sequence, the missile could be raised and fired. A nuclear artillery regiment included three batteries for a total of six AMX30Ps. Each of the five nuclear artillery regiments also maintained a reserve of two AMX30P vehicles, which permitted maintenance

BELOW This is the rear of the launch container, which was equipped to support the missile's rocket component during transportation and as a launcher unit. (Fonds Claude Dubarry, Collection du Musée des Blindées de Saumur)

LEFT Once the launch container was carefully installed on the launcher platform, the crew called up the second lorry, carrying the unarmed warhead. The Pluton battery aimed and controlled missile fire with the IRIS 35M fire control computer, which incorporated an operator screen, printer and a set of modems to communicate within its battery control network. The IRIS 35M was operated by the *operateur-controlleur*. The AMX30P's auxiliary power plant (GAP) allowed a constant supply of power to the IRIS 35M fire control computer and the VP213 wireless system, which was employed to maintain communication with the battery command post. As a result the AMX30P could operate its electronics without running the main engine if necessary. *(Fonds Claude Dubarry, Collection du Musée des Blindées de Saumur)*

BELOW Installation of the Pluton's unarmed warhead is completed. The Pluton's warhead received its nuclear core prior to firing, and was only armed by the crew at the last minute. In flight, the Pluton was guided to its target by an inertial guidance mechanism. The aiming was accomplished with the IRIS 35M system – a smaller version of the IRIS 50 fire control computer developed in 1967 for use in artillery batteries. As fitted to the AMX30P, it was exceedingly simple in function compared to a modern computer, serving to calculate trajectory and fire control parameters. It had a memory of 32 Ko, miniscule by modern standards, but adequate to assure communications with the battery control's own 16 Ko IRIS computers, synchronisation of a fire plan and also to serve (in test function) as a launch simulator for training. Inside the AMX30P the IRIS 35M installation consisted of the computer unit, the operator input keyboard and screen and of a series of primitive modems (modem was the abbreviation for *modulateur–démodulateur,* the data exchange device which tied the IRIS 35M to the rest of the TVRM15 battery communication network, like a simplified version of a computer modem as we know it). *(Fonds Claude Dubarry, Collection du Musée des Blindées de Saumur)*

RIGHT The communication network employed within each artillery regiment included wired and wireless transmission equipment, with relay stations (usually mounted in heavy communication lorries) to ensure signal strength. The IRIS 35M system functioned with the onboard Pluton missile control system (known as the SMOC or *système de mise en œuvre et de contrôle*) which controlled commands, data input for launch conditions and elevation commands to the hydraulic system, alignment of the missile gyroscope, temperature controls and missile onboard battery charge. Like the Roland and AU F1, the AMX30P's electronics added substantially to the vehicle's unit price and Thomson-CSF and other electronics companies were substantially involved alongside SNIAS and GIAT (both in development and system integration, and subsequently over the life cycle of the weapon system). *(Fonds Claude Dubarry, Collection du Musée des Blindées de Saumur)*

and replacement as required. A total of 40 AMX30P production vehicles were constructed at ARE. These were in active service between 1974 and 1993.

AU F1 self-propelled gun

One of the most famous AMX30 variants was the *Automoteur F1*, or AU F1. This 155mm self-propelled gun system consisting of the 155mm *Grande Cadence de Tir* (high rate of fire or more usually 'GCT') turret mounted on a modified AMX30 hull. Some 273 of these weapons, including prototypes, were manufactured at ARE and Bourges for the French artillery and at least 153 were manufactured for export. When the first AU F1 entered service in the French artillery in 1980,

RIGHT W590-102, the first prototype of the AMX30 155 GCT, was delivered to the STAT on 1 August 1973. The GCT's 39-calibre 155mm gun could fire 155mm hollow-base shells to a 23km range – with an impressive rate of fire of six rounds in 45 seconds. It could also fire a variety of older ammunition types, usually to a range of 18km. *(Archives CAAPC)*

The massive doors in the turret rear lowered into platforms that permitted rapid reloading. The GCT's 42-round automatic loader represented the cutting edge of artillery technology in the early 1970s. The width of the original turret was greater than the railway loading gauge, forcing a redesign which permitted the GCT to be carried by rail along standard SNCF lines. *(Archives CAAPC)*

ABOVE A maximum elevation of 66 degrees was possible for high-angle or short-range fire. The chassis was reinforced not only to counter the weight of the turret but also to mitigate the effects of heavy recoil (and the repetitive shock of a high rate of fire). The special heavy-duty road wheels, also seen on the AMX30H bridge-layer were the only outwardly visible trace of the reinforced suspension system originally envisaged. These special road wheels were not retained for the AU F1 production vehicles. *(Archives CAAPC)*

BELOW In 1974 six prototype AMX30 155 GCT self-propelled guns were evaluated by the *40e Régiment d'Artillerie*. The number was sufficient for a complete battery with a single replacement vehicle. Among these were two guns fitted with fume extractor sleeves, including *La Marne* seen here reloading, and 694-0158 *Verdun*. This configuration was not retained for the production AU F1, which used a compressed air evacuation system to purge fumes from the barrel. The average replenishment time for two men was around 20 minutes, and firing could continue during the reloading operation. *(Archives CAAPC)*

it was the only self-propelled gun in any NATO army with an automatic loading system, and it was possibly the most advanced self-propelled gun system in service in any army.

The GCT concept was devised at the EFAB Bourges between 1967 and 1969. The original concept was to create a self-propelled weapon for divisional artillery regiments which could fire top-attack munitions or shells containing anti-armour sub-munitions. The idea was to fire shell-delivered cluster mines which could deny territory to an attacking force or which could disrupt mechanised attacks by penetrating the roof armour of enemy main battle tanks. The top-attack sub-munition was designated the TIAB (*Tir Indirect Anti Blindée* or Indirect Anti-Tank Fire) programme. As the TIAB munition was expected to be employed by divisional artillery, the study of a 155mm gun design suitable for firing these munitions was initiated in 1968. The *Arsenal de Bourges* design team came to the conclusion that in order to effectively fire TIAB munitions the artillery system would need to be highly mobile and to possess a higher rate of fire than a normal 155mm divisional gun. A rate of fire of six rounds in 45 seconds was specified, and this ultimately was achieved in the production GCT turret.

Such a high rate of fire with a turret-mounted medium-calibre gun firing rounds consisting of a separate shell and bagged charge and loaded by human loaders proved impossible. As a result, development of a 39-calibre 155mm gun served by an automatic loading system began at Bourges. The 155mm gun design drew on existing French design experience for its barrel and muzzle brake and it used a breech equipped with obturator plates. The whole TIAB concept had run into numerous practical problems by 1971. The biggest issue lay with

LEFT In 1974, the 'AMX30 155 AU GCT' received its official designation '155 AU F1'. Despite being a bit heavier than the AMX30 (42 tonnes in combat order) and sharing the same engine, the cross-country mobility of the gun was comparable to the tank, as was the maximum speed on roads (60km/h). *(Archives CAAPC)*

155mm ammunition types used widely in the NATO armies or older French ammunition types such as the OE Mle 56 (*Obus Explosif Mle 56* adopted in large quantity by the French Army and manufactured from 1956). The CN-155-F1 fired the OE Mle 56 at a velocity of 790m/sec to a range of 22km. This capability made economic sense, and a self-propelled gun using the CN-155-F1 could use the substantial existing stocks of French 155mm ammunition fired by the AM F3 155mm gun or existing towed 155mm guns. The CN-155-GCT's optimum performance came when firing the new 155mm OECC F1 round (*Obus Explosif a Culot Creux F1*, or hollow-based high-explosive shell model F1). The OECC F1 was a 155mm round first developed in 1960, but it had not yet been produced in quantity owing to the lack of a gun to fire it optimally (and it was produced alongside the AU F1 from 1975 onwards). The CN-155-F1 fired the OECC F1 round at a velocity of 810m/sec to a maximum range of 23.5km.

BELOW The AU F1 stood an imposing 3.17m tall without the loader's 12.7mm machine gun. Taking into account the limitations imposed by the rail transportation regulations, a new type of low-bed flatcar had to be designed to carry the new self-propelled gun. *(STAT)*

the limitations of the 155mm round itself. A medium-calibre artillery shell was not large enough to carry a worthwhile number of sub-munitions, and this led to the pursuit of area denial/top-attack munitions delivered by rocket artillery instead. The 155mm gun and the GCT turret system designed for it were reprioritised to deliver high rate of fire divisional artillery support with conventional 155mm munitions.

The 155mm gun itself was designated CN-155-F1 by the French Army and as the *Canon de 155mm Grande Cadence de Tir* (or CN-155-GCT) within GIAT. The gun was versatile enough to easily fire American-manufactured

The turret designed for the CN-155-F1 and its automatic loading system by the EFAB Bourges was designed and built in 1969–70, designated *Tourelle 155 GCT* – or simply GCT. The GCT's automatic loading system could handle the older projectile types described, or US-produced projectiles. In addition to high-explosive shells, the GCT system could fire illuminating, smoke and practice rounds. The OECC F1 round in particular could be equipped with a variety of fuses, most notably with the FURA F3 type which could be set by the loader within the confines of the GCT turret. This percussion fuse could explode at 12m height, or on impact. Although the GCT system was intended for use as an indirect-fire weapon, it was provided with sights to permit its use as a direct-fire weapon if necessary.

Typical 155mm and similar medium guns fired a separate shell and bagged charge loaded by hand, the breech being sealed by obturator plates. In the case of the CN-155mm AU F1, the automatic loading system required a rigid combustible case, formed from a stout nitrocellulose-impregnated kraft paper casing (essentially a cardboard tube weighing 1.8kg). Development of the rigid combustible charge began at the *Service de Poudres* research establishment at Bergerac in 1969 and the first test-firing of the 39-calibre 155mm gun with the combustible case first took place in 1970

at Bourges. The technology adopted for the combustible case ammunition was unique for its time, and it was developed for maximum range with the hollow-based OECC F1 round. It was also compatible with the conventional 155mm rounds, and use of these did not affect the rate of fire possible with the automatic loading system.

The rigid combustible casing contained a series of smokeless powder charges in silk bags, differing in seven different weights from light to heavy. The casings were issued pre-loaded to maximum range, but by removing a stopper the *Radio/Artificier/Chargeur* could reduce the charges incrementally to cut down the range to

ABOVE **The Uais flatcar's design allowed the passage of the 155mm gun's muzzle brake during the downward movement of the AU F1.**
(Archives CAAPC)

LEFT **Fording capability tests were part of the AU F1's commissioning test at Bourges. The AU F1 could ford to a depth of 2.10m according to the technical manual, but lacked the deep-wading capacity of the AMX30B.** *(STAT)*

ABOVE The overall *vert OTAN* (NATO green) paint scheme worn here by the *Souville* was typical for AU F1's before 1984 (although some vehicles retained this scheme until 1989). Here we can see the standard gun fume evacuation system. The crew's personal kit was stowed on a dedicated rack located on the rear of the turret. *(STAT)*

BELOW A selection of the large range of French 155mm ammunition compatible with the CN-155-F1, including a combustible casing in the foreground. There are two OE-OCC-155-F1 rounds on the far right, an older OE Mle 56 round in the middle, an OF-OCC-155-F1 smoke round in pale green and a blue OF Mle 56 smoke round. *(Olivier Carneau)*

specified distances. The charge bags weighed as follows: No 1: 0.88kg, No 2: 1,45kg, No 3: 4.4kg, No 4: 5.6kg, No 5 8.3kg, No 6: 10kg, No 7: 11kg. Firing was accomplished by means of an electrical circuit which ignited a 50g primer and 250g primary charge in the base of the combustible case tube.

The GCT turret itself was built of 10–20mm welded armour steel plate, giving adequate protection from small arms and shell splinters. The turret sides were each equipped with a rearward-hinged crew door and the roof incorporated a single hatchway on the loader-operator's side equipped with a machine-gun mounting for local defence. The commander was provided with an octagonal cupola (described by the artillery as the *kioske*) fitted with eight episcopes for battlefield surveillance. The fire controls were conventional, having been developed from the type first proposed for the AMX13-based 105mm turreted self-propelled gun developed for the Swiss Army in 1963. The optical devices included a periscopic gunnery sight with provision for direct fire. The automatic loading system occupied the entire rear of the turret.

In 1970 the first prototype GCT turret was completed, and it was shown that same year temporarily mounted on a standard AMX30 tank hull. Without an auxiliary engine to

power the electrical and hydraulic systems of the GCT turret in such a configuration, the driver had to keep the engine running constantly (or the batteries would drain in an unacceptably quick time). This arrangement proved that the hull required reinforcement, the suspension required thicker torsion bars and the provision of an auxiliary motor to power the turret's hydraulics and electrical system was an absolute necessity. An AMX30 hull was adapted comprehensively, to include the installation of a Citroën 4kW auxiliary petrol engine (GAP) in the right-hand side of the glacis. Its exhaust system extended on to the right sponson in order to divert fumes from the front of the vehicle. When the self-propelled gun was in action, the GAP's cover plate had to be removed and was clamped to the left-hand side of the turret. The first prototype was delivered to the army in 1971, and extensive evaluation followed over the course of the next year. Trials revealed a number of necessary changes required to the basic turret design as well as to the chassis, but were overall very successful. An order for the production of a *preseries* was issued in 1973.

In early 1974 six redesigned 'AU AMX30 155mm GCT' *preseries* vehicles were delivered to the French artillery for field testing with the *3e Batterie, 40e Régiment d'Artillerie*. These

vehicles served to establish five-gun battery tactics for the creation of full regiments of 20 guns. The first order of 12 production AU F1s (the designation having changed at the time of the first order) was placed later the same year. Turret manufacture was undertaken at Bourges, while the hulls were constructed on the AMX30 assembly line at ARE. The EFAB Bourges built special test facilities, which included 30km-worth of roadways to evaluate mechanical performance, a 2.1m-deep concrete wading tank for 5-minute wading seal tests with the motor running and a 30-minunte cross-country test-track. A test-firing range was also built on the site.

One of the problems associated with the AU F1 was its very high unit cost. This came in part from the use of the automatic loading system, but also because the French artillery intended to equip the AU F1 with instrumentation that would permit it to fire accurately under any conditions. An onboard gyroscope, goniometer and the ATILA artillery fire control computer all added substantially to its unit price. The cost of the AU F1 was a problem that GIAT hoped to mitigate with foreign sales. As a result a significant effort was made to demonstrate the gun system to the Royal Saudi Arabian Army, which placed an order for 63 systems. These guns were officially

ABOVE The main armament was hydraulically controlled by means of a motor for traverse (at 10 degrees per second) and two cylinders for elevation (at 5 degrees per second). These cylinders also served to stabilise the gun. The new double-pin 'live' tracks indicate that this vehicle was upgraded during the late 1980s, very likely with the new *microturbine* GAP at the same time. *(STAT)*

designated AMX30 155 AU within GIAT and for foreign sales, and deliveries were made ahead of production for the French Army. Deliveries for an Iraqi order for 83 more followed in 1980.

The GCT turret

It is believed that between 427 and 430 155mm GCT turrets were built by the EFAB Bourges, including prototypes. The turret retained its GCT designation throughout its production for the AU F1. The automatic loading system was designed to fire six rounds in 45 seconds, and as a result the ammunition was held in columns of six projectiles and six charges respectively on each side of the breech. Two loading arms were attached to a breech-mounted trackway extending the width of the turret. These fed the breech and functioned as a rammer. The GCT turret's loading system worked from the two innermost racks outwards (in sets of six rounds). The left-hand arm loaded the shell from the left side of the breech, ramming it into place, followed by the right-side arm which fed in the casing and pushed it clear of the obturators. The breech then closed and sealed, and the gun was fired. Each rack in the automatic loading system fed from bottom to top, moving outward by column as the magazine was emptied.

On the prototype GCT turrets the system featured seven racks on each side for a total of

42 rounds, but after user trials it was necessary to modify the width of the turret roof in order to clear the existing railway tunnels. On all of the production turrets the outermost shell rack and the outermost charge tube rack were not accessible to the automatic loader, serving as ready racks for reloading. The production GCT turret could be identified by the rounded corners on the roof plates. A total of 36 rounds were available to the automatic loader with six reloads available in the outermost racks.

Production versions of the AU F1

The AU F1 underwent substantial improvement during its service life. The only upgrade dedicated to the AMX30-based hull of the AU F1 was implemented in 1990. This involved the substitution of the Citroën GAP with a far more powerful *microturbine* 12kW GAP manufactured by Gevaudan. At the same time an improved fire control system was added to the 155mm GCT turret, known as the *conduit de tir inertiel* (inertial fire control system). The modified vehicles received the designation AU F1 T, while the original production version of the production gun was redesignated AU F1 H at the same time. AU F1 Ts included vehicles remanufactured from standard AU F1 H vehicles and some may have been completed to the later standard from the Bourges and ARE assembly lines in 1990–91.

These were followed in 1991 by the last version produced on the original AU F1-pattern hull. The last 24 vehicles produced incorporated further improvements to the fire controls, and were designated AU F1 TM (*Tourelle Modex* or *Tourelle Mode Experimental*). The AU F1 TM incorporated updated turret electronics and communications systems to test out the new ATLAS artillery communications system. The only exterior difference visible to distinguish these from AU F1 T production vehicles was the provision of a second antenna mount on the turret roof next to the commander's cupola. The original antenna mount was sealed and was not employed with the vehicle's communications system. The left side of the gun mantlet incorporated an electrical outlet to power a gun-mounted cinemometer, measuring the velocity of the projectile. The 24 AU F1 TMs were all issued to the *40e Régiment d'Artillerie* to form a complete regiment of three eight-gun batteries.

The AMX30 155 GCT AU was used in combat by the Iraqi Army against Iran with successful results in the 1980s. Saudi GCTs

ABOVE In June 1996 NATO deployed troops into Bosnia in order to dissuade the warring factions from attacking one another. In August 1996 an eight-gun battery of AU F1 Ts was detached from the 40e RA as the *Groupe d'Artillerie Leclerc* and was deployed to Bosnia as part of the NATO Rapid Reaction Force. The battery was tasked to provide fire support to French forces assigned to NATO forces peacekeeping in the area. On 20 August 1996 the AU F1 Ts took up positions with British and Dutch Army units on Mount Igman, west of Sarajevo. On 30 August the battery conducted Operation Vulcain, a bombardment of suspected Serbian artillery positions. Within 24 hours Serbian Army command posts and assembly areas around Lukavica were bombarded with precision. The eight guns fired over 300 rounds within the first 48 hours in a carefully surveyed bombardment on pinpoint targets, mindful of possible civilian losses. The Serbian positions were hit badly enough to force a withdrawal. *(US NARA)*

RIGHT The auxiliary power unit was upgraded to the 12kW *microturbine* in 1990 on a substantial number of AU F1s, which led to the change in nomenclature to AU F1 T. This adaptation was also seen on the 24 AU F1 TMs. *(Zurich 2RD)*

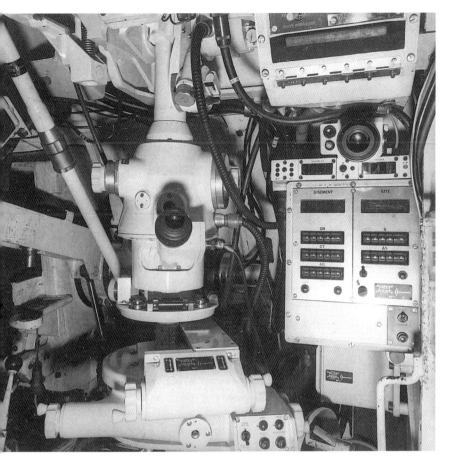

were employed during Operation Desert Storm but very few details have been made public. In 1991 18 AU F1s were delivered to the Kuwaiti Army, too late to serve in the liberation of their country. France employed the AU F1 T with great effectiveness against Serbian forces in August 1995 from Mount Igman near Sarajevo. In all cases, the high rate of fire possible with the GCT turret's automatic loading system and the long range of the 39-calibre CN-155-F1 gun proved the AU F1 as an exceptional weapon system.

The AU F1 was expected to be replaced (or possibly remanufactured) into the AU F2 around 1995, but the end of the Cold War reduced the urgency of this programme as the army juggled its priorities, which included the adoption of new weapons like the Leclerc battle tank. The AU F2 would have incorporated a 52-calibre-long 155mm gun, new munitions permitting a range of over 40km, ultramodern communication systems and a host of other improvements. Its hull might have been adapted from the AMX40 hull or even from the Leclerc itself. The AU F2 was delayed repeatedly and was eventually cancelled in the mid-1990s. The army decided to modernise 104 existing AU F1 turrets with the ATLAS artillery fire control system, new communications equipment and rebuilt 39-calibre guns. These turrets would be installed on rebuilt AMX30B2 tank hulls suitably modified for the stress and weight of artillery operations. This series was converted in 2003–4 at ARE and designated AU F1 TA (TA for *Tourelle Atlas* or Atlas Turret). With its introduction, all earlier AU F1s were withdrawn from service by the end of 2006.

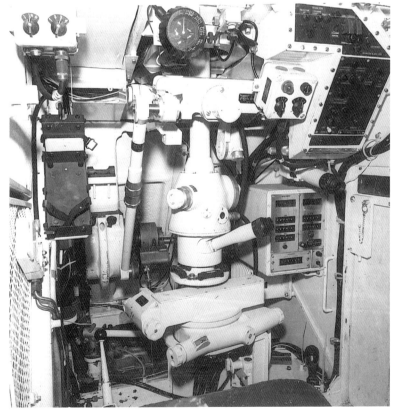

The AU F1 TA was an easier vehicle to drive with its automatic transmission, and it enjoyed a slightly better cross-country performance than the AU F1 H or AU F1 T thanks to the new Mack E9 engine. The AMX30B2 hulls employed in the conversion were mostly early-pattern vehicles built between 1984 and 1987. The hulls retained many of the exterior features typical to the AMX30B2 battle tank hull and the main changes made to the AMX30B2 hull as seen from the outside were confined to the addition of stowage and exhaust covers per artillery regulations. Distinctions from the previous AU F1 hulls were numerous and these included the AMX30B2-style glacis (with the GAP omitted completely). The turret ring was raised 12cm in order to clear the engine deck (raised slightly from the standard AMX30B2 engine deck for the E9 engine conversion). The AU F1 TA's turret could be distinguished from the exterior by its two roof-mounted antennae (the position of its original AU F1 H antenna mount being welded over but still visible on the roof) and the electrical plug outlet on the left side of the mantlet. While the AU F1 TA theoretically lacked the sort of mobility of the

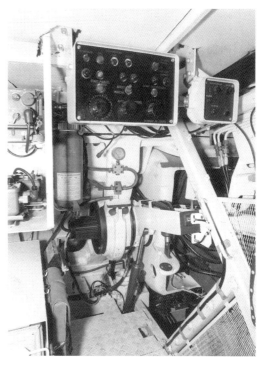

LEFT The radio set is partly visible on the left side, the loader being also responsible for the communications. With each successive version of the AU F1, progressively more sophisticated equipment was introduced. *(Archives CAAPC)*

BELOW *Herbsheim 1945* was an AU F1 TA photographed during the preparatory phase of the Bastille Day parade in Paris. The donor AMX30B2 hull used in this AU F1 TA conversion was delivered as a battle tank in 1986. The AU F1 TA was the most sophisticated version of the AU F1, equipped with the ATLAS artillery fire control system. *(Jerome Hadacek)*

ABOVE Four AU F1
TAs were deployed to
the UNIFIL contingent
between 2006 and
2011, where they
served to dissuade
artillery attacks on UN
forces. *(Olivier Carneau)*

Leclerc MBT, which had by then replaced the
AMX30B2, it proved to be a very serviceable
conversion. Its crews certainly regretted that the
microturbine GAP could not be incorporated
into the AU F1 TA – but with this exception it
was the most capable production version of the
self-propelled gun.

The AU F1's crew

The AU F1's driver operated the vehicle very
much in the same manner as the AMX30B,
with its manual transmission and tiller bars. He
would have struggled to exit the vehicle with
the turret traversed anywhere but forwards.
The GAP mounted next to him gave off a
substantial amount of heat during operation, as

well as exhaust fumes. The gun crew sat in the
turret in front of the automatic loading system's
magazine, caged off from the breech while the
gun was in operation by a system of wire mesh
and perforated metal guards. The automatic
loading system operated behind their positions
on either side of the breech. The *chef de pièce*,
or gun commander, sat to the right-hand side
of the breech, with the *pointeur* (gun layer) to
his immediate right. They could, if necessary,
share the gun-laying controls and sights, while
the commander was also provided with an
interface for the ATILA battery control system
or for the ATLAS gun computer (depending on
the version) at his position. The *Radio/Chargeur/
Artificier* (RCA, or loader) was seated to the left

RIGHT This is one
configuration that would
have been employed for
the AU F2,
a weapon system
developed from the
AU F1's GCT turret allied
to one of the four AMX40
prototype chassis
(which might have been
replaced with a Leclerc-
derived chassis on a
production vehicle).
The turret included a
52-calibre version of
the CN-155-GCT, with a
range surpassing 40km.
It would have included
the ATLAS fire control
system, and possibly a
battlefield management
system. Its cancellation
was but one of many
casualties suffered
by GIAT after the
Cold War ended.
(Jerome Hadacek)

of the gun breech, isolated by the breech guards during firing and dependent on the intercom for instructions from the vehicle commander. The loader was responsible for operating the automatic loading system. He also set fuses and prepared the ammunition – and operated the wireless system. The loader was also tasked with the local defence of the vehicle. His roof hatch ring was equipped with a 12.7mm heavy machine gun (a licence-built M2 HB Browning).

The AU F1 was something of a technical marvel with its inertial navigation and sophisticated gunnery aids. The French artillery arm invested heavily in integrating its artillery systems into a battlefield management system as a standard doctrine starting in the late 1960s. Artillery regiments were organised with radar-guided fire control centres, and an intricate system of communications for battery command and control at the divisional level. The provision of an inertial navigation system was considered a key element in this, and onboard inertial navigation capability was an integral part of the GCT concept from its early stages. While this type of equipment (based on gyroscopic equipment) had been under development for the French artillery since 1968, it was not fully perfected as an onboard system for the GCT turret until 1975. The inertial guidance and battery communication systems were expected to provide a means of centralised fire control,

rather than to give the individual gun commander increased initiative in target selection.

The AU F1 could be equipped with a miniature Doppler radar (MIRADOP) mounted on the gun tube to aid in fire control. The ATILA system was employed to communicate directly with the battery command vehicle and functioned as a primitive form of network communication. Exchanges of information took roughly 60 seconds – a miracle of technology for the early 1980s. The ATLAS system that replaced ATILA at the end of the 1990s was far more capable, permitting the gun battery to function as part of a battlefield management system, expanding the range of battlefield tasks possible with an artillery battery.

At the time of its introduction the AU F1 regiment deployed three and then four batteries of five guns. In the early 1990s the reduction of the artillery arm resulted in the creation of the eight-gun battery organisation, each of which could separate into two four-gun *sections de tir*. With a regiment of AU F1s controlled from mobile fire command centres, 32 of these guns could fire 6 rounds apiece into a target the size of a football pitch in under a minute. It was just the kind of capability which might have broken up an armoured attack, destroyed a vital bridge crossing, a crossroads or an enemy artillery battery.

The GCT turret was intended to be a suitable

LEFT The GCT turret was never successfully sold to any of the NATO armies despite its compatibility with most 155mm ammunition types. This was most likely due to the high unit cost of the turrets in comparison with the American M109 series of self-propelled guns. This Leopard GCT trials vehicle was tested in 1973. (*Fonds Claude Dubarry, Collection du Musée des Blindées de Saumur*)

match to a range of standard hulls, particularly that of the Leopard. As a result, a contract was signed between GIAT and Krauss-Maffei to test out the Leopard with a GCT turret in 1973, in the hope that France's allies would buy the admittedly expensive weapon system for their own artillery arms. Even 20 years after the type entered production, the possibility of adapting the GCT turret to the T-72 M1 hull and to the Arjun Main Battle Tank for the Indian Army was evaluated by GIAT Industries, albeit without success.

AMX30 Roland and Shahine

In 1963 talks began between the French and West German governments to co-develop and co-produce a surface-to-air missile, which developed into the Roland 1 and Roland 2 missile systems. The clear weather Roland 1 was taken on as the responsibility of the French government and the Roland 2 (an all-weather version of the same missile whose development was expected to take 24 months longer) was taken on by the West Germans. Parts for the missiles were produced in both countries, and

the West Germans also adopted the French Roland turret design as the basis of the turret for launching the Roland 2 missile (as did the United States). When the Franco-German partnership to co-produce the missile was agreed, a company was formed from elements of Nord Aviation, Thomson-CSF and Messerschmitt-Bölkow-Blohm, under the name Euromissile. Over 23,000 missiles were manufactured before production ceased in 1989.

The AMX13 chassis had already proved too small for a surface-to-air missile launcher, but proved the basic concept of radar target acquisition and visual engagement in clear weather, and resulted in the requirement for a prototype AMX30-based Roland 1 launcher. The modifications to turn the basic AMX30 hull into a suitable basis for a guided missile-launching vehicle (known as the AMX30R) were substantial. A tall, boxy superstructure replaced the driving and fighting compartments with seats for the crew of three abreast behind the front glacis.

The Roland turret was equipped with twin elevating firing rails. The rear of the turret provided the mounting for a rotating target acquisition radar. The hydraulically operated

RIGHT The potential of the AMX30 as a suitable chassis to serve as a platform for air defence artillery (missile and gun) was recognised very early in the programme. This **AMX30R Roland 1** was one of the first ones built, and was photographed while under evaluation at the STAT in the early 1970s. *(STAT)*

Roland turret was perfected as a joint effort between *Aérospatiale*, GIAT, Thomson-CSF and SAMM. West German-produced components were also employed, including the launch rails and associated equipment. The Roland 1 and Roland 2 missiles were both adopted by the French Army and the turrets designed to fire each differed considerably. The Roland 1 missile could be fired by either turret. The Roland 1 turret was equipped with the three-channel Thomson-CSF MTI search and tracking radar mounted on the turret roof, but the system depended on the missile operator keeping the target in the sight reticle up to the point of detonation. A single AMX30R prototype (bearing the serial 624-0204) was constructed in early 1972 and delivered to the STAT on 22 March 1972 (with simultaneous provision of a Roland 1 turret to West Germany to mount on a Marder Schutzenpanzer). This was followed by three AMX30R Roland 1 pre-production vehicles delivered in 1974. The national assembly voted through funding for an order of 20 production AMX30R Roland 1s in the same year.

The first production batch of AMX30R Roland 1 was produced and delivered in 1977. Funding problems in the mid-1970s, rather

ABOVE The turret returned to the 12 o'clock position with firing rails lowered when travelling or when reloading. In this case the machine is at rest, with its radar folded down. The missile magazines were located on each side of the turret. *(STAT)*

than technical problems, delayed the broader adoption of the AMX30R. As a result the type entered service as partial equipment in several regiments, where it served alongside older equipment. In all 174 Roland turrets were completed for mounting on French AMX30R

LEFT An AMX30R Roland 2. The Roland was a tube-launched missile system. Each missile was 2,400mm in length, with a diameter of 160mm and a wingspan of 500mm. Each missile weighed 67.8kg in its firing tube, which was ejected after firing. The missile alone weighed 66.5kg. The Roland 1 missile fired at a velocity of 500m/s, propelled by a two-stage rocket booster. The Roland 1 was effective to an altitude of 4,500m, the Roland 2 to 6,000m. The missile was controlled by a command microwave radio link. The Thomson-CSF Dassault radar was effective to 16km. *(STAT)*

ABOVE The Roland 2 turret was the all-weather version equipped with a second two-channel Siemens radar mounted on the turret front. The second radar system permitted the Roland 2 missile to be kept on target once designated by the operator, automatically following the target regardless of weather conditions. *(STAT)*

LEFT In the AMX30R1 and AMX30R2 the turret was manned in action by the missile operator, but the unit was slaved to the complex fire controls and provided with an exceptionally rapid traverse. If a target was identified with the vehicle halted, it could fire within 15 seconds. In order to reload from its twin four-missile revolver-type magazines, the turret returned to the 12 o'clock position automatically and the missile firing rails lowered to reload. This process took less than 10 seconds, and it was repeated after each missile launch. Loading the empty magazines was a three-man operation, with special ladders that hooked on to the hull sides. *(STAT)*

Roland 1 and Roland 2 launcher vehicles. This total included 81 AMX30 Roland 1 types delivered between 1978 and 1981, and 93 Roland 2 turrets delivered from 1981 until 1987.

Normally each AMX30R was accompanied by a VAB (*véhicule de l'avant blindée*) wheeled APC that carried a second crew to permit maximum vehicle availability. Power for the complex electronics, high-power hydraulics and to keep the vehicle batteries consistently charged was provided by a 60kW auxiliary engine, which was integrated to operate with a specially produced version of the HS-110 engine (which was not interchangeable with the standard version employed in the AMX30B and was not compatible with transmission upgrades). The turret basket was guarded to prevent injuring the crew during operation. Training with the Roland equipment could be accomplished through a dedicated simulator that turned the AMX30R itself into a training system through an electronic link. When the army was reduced at the end of the Cold War the French anti-aircraft artillery regiments retired the AMX30R Roland 1 very quickly, standardising on the AMX30R Roland 2 before this system too was withdrawn from service in 2008, to be replaced by the Mistral system.

AMX30 Shahine

A more capable, and far more costly surface-to-air missile system was also developed for export to Saudi Arabia in the form of the SA-10 (R460) Shahine. This system was developed into a self-propelled system based on the AMX30 chassis

in the mid-1980s. The SA-10 (R460) Shahine missile was an enlarged, improved, two-stage version of the successful Thomson-CSF Crotale missile. The Shahine was a medium-range SAM system, requiring a more powerful radar and heavier missile mounting. The design of separate radar control vehicles and missile launch vehicles was necessary by virtue of the sheer size and bulk of the equipment. As such, the Shahine weapon system was built around the battery rather than as a self-contained radar and launcher vehicle like the Roland.

Thomson-CSF and GIAT collaborated on the manufacture of the Shahine AMX30SA *Char de Surveillance* (which might be more accurately described in English as a radar tracking vehicle), which mounted a powerful pulse Doppler search radar system. The basic hull was based on the Roland hull but the larger radar unit required the crew to mount through a large access hatch in the front of the hull. Each *Char de Surveillance* acquired targets for two AMX30 *Char Missile* Shahine launcher vehicles. These were based on a similar pattern hull to the *Char de Surveillance,* but mounted a massive traversable six-missile launcher with a frontally mounted tracking radar. Reloading the SA-10

ABOVE The rear of the AMX30R was laid out very differently from the battle tanks. It incorporated an auxiliary motor to power the electrical system, and employed a special version of the HS-110 engine. *(Jerome Hadacek)*

BELOW The AMX30 Shahine *Char Missile*. *(Jerome Hadacek)*

missiles was accomplished externally with a lorry-mounted crane, which lifted complete firing tubes on to the launch unit. It is believed that 12 complete batteries were sold to Saudi Arabia, but much regarding the Shahine system remains classified.

AMX30 Bitube 30mm DCA

The determination to equip the French artillery with the most effective anti-aircraft weapons possible was a priority in the post-war army budgets, but development of such weapons could only advance at the pace of its available funding. After extensively testing a range of captured German anti-aircraft weapons at Bourges in 1945 and into the 1950s, DEFA's engineers settled upon twin 30mm automatic cannon as the optimal mobile anti-aircraft weapon to protect mechanised units from low-level air attack. The French SAMM (*Société pour les Applications des Machines Motrices*) company was tasked with developing a radar-directed 30mm dual automatic cannon turret system with a rate of fire of 600 rounds per minute. French efforts to develop a self-propelled anti-aircraft gun turret based on twin 30mm Hispano-Suiza HS831 automatic cannon armament received US Military Aid Program funding for 25% of all development costs in 1957.

The SAMM turret was designated Type S401A, and it employed hydraulic pressure to traverse 80 degrees per second, tracking at 45 degrees per second. It carried a total of 150 rounds per gun, and it was mounted on the AMX13 turreted 105mm self-propelled gun chassis. The S401A turret was retrospectively fitted with the *Oeil Noir* radar by the Marcel Dassault company (which could be folded into a box fitted to the turret rear when not in use) prior to adoption. Funding issues delayed the S401A's entry into service. SAMM delivered the turrets in 1963 without their radars, or fire controls. The combined search and tracking radar was ready to enter service by 1966. The turrets remained in storage unissued for another two years. Tests of an S401A turret (without radar) mounted on AMX30A prototype 234-0287 were undertaken in 1964 alongside the AMX13 Bitube 30mm DCA.

These trials showed that the larger chassis could carry the turret without any impact on cross-country performance and that the AMX30 chassis could carry twice as much ammunition – some 300 rounds per gun. Using the AMX30 chassis also demanded the provision of an auxiliary engine, but the French Army had already ordered the lighter type and had

LEFT The AMX30 Bitube DCA's TG230A turret was essentially the extrapolation of the AMX13 Bitube 30mm DCA's S401B turret. The TG203A was designed from the outset to include the improved *Oeil Vert* radar and with the benefit of the experience of developing the AMX30R Roland 1 missile fire control system. The AMX30 Bitube DCA was built for the Royal Saudi Army, entering service in 1979. *(Thomson-CSF)*

committed to the development of the SABA missile. The AMX13 30mm Bitube DCA was not available to the French artillery as an integrated, operational weapon system until 1969 – about 12 years after development started, and long after the Roland missile programme was instituted. By this time the French Army was convinced of the superiority of surface-to-air missile defence for its mechanised units.

The S401A turret was ill suited to the small AMX13 chassis, which limited ammunition capacity and suffered from a poor cross-country performance. The AMX30 Bitube DCA concept was improved considerably by mounting an updated version of the S401 turret, designated the TG230A. The TG230A turret was fitted with the Thomson-CSF *Oeil Vert* radar derived from the Roland system and twin SAGEM direct fire sights for use against ground or low-level air targets. The turret was mounted on a lightened version of the AMX chassis designed for the GCT 155mm self-propelled gun. This chassis included an auxiliary engine to power the turret's radar system and hydraulics, and it had room for 900 rounds of 30mm ammunition. In 1974 when this design was finalised the French Army was testing the AMX30R and the army evaluated the self-propelled anti-aircraft gun (SPAAG) system

BELOW The French Army never purchased the AMX30 Bitube DCA, having introduced the smaller AMX13 Bitube 30mm DCA in 1969 after multiple delays. AMX30A 234-0287 was tested out with an S401B turret to prove the concept in 1964–65, being refitted with its 105mm turret following the trials. It was no doubt as a result of these trials that the production vehicle was built on the same basic hull type as the AMX30 Au 155 (GCT) with its auxiliary motor, seen here on the lower left edge of the photo. The turret traverse and gun elevation required a powerful hydraulic pump, and the radar system's electronics used a substantial amount of power. The auxiliary motor was a Citroën 4kW unit. *(Thomson-CSF)*

manufactured, one of which was retained as a reference vehicle by GIAT.

Buoyed by the successful sale of the AMX30SA, Thomson-CSF and GIAT collaborated in a subsequent attempt to produce a viable 30mm anti-aircraft turret, as a modular weapon system in the spirit of the GCT turret. The system was marketed not only with GIAT support but also with the British Royal Ordnance Chieftain in mind. A modernised twin 30mm HS turret with updated *Oeil Vert* radar marketed by Thomson-CSF as the *Sabre* was trialled on an AU F1 chassis in the early 1980s. The radar functioned in a similar manner to the type employed on the Roland 1 turret and was effective in clear weather. The turret was characterised by its lightly armoured welded shell with numerous angles to permit as much space as possible inside the turret. The *Sabre* was also demonstrated on a modified AMX10RC wheeled chassis. The problem of rapid barrel wear was a serious issue that affected further development of the Hispano-Suiza 30mm weapon. The *Sabre* came just as most of NATO's armies were investing in surface-to-air missiles and proved unable to secure any orders despite its suitability for mounting on a wide range of tracked and wheeled chassis.

ABOVE A modernised, modular version of the TG230A turret was marketed as the Thomson-CSF *Sabre* system in the early 1980s. This could be mounted on a range of different hulls, and it was tested out on the AMX30 Au 155 (GCT) type hull, the AMX10RC armoured car and the British Chieftain tank hull.
(Jerome Hadacek)

RIGHT The *Sabre* was marketed heavily with GIAT's Middle Eastern clients. The main drawback with the *Sabre* was that its gun barrels wore very quickly, and while it might have proved possible to employ an alternative gun, no orders were ever placed.
(Jerome Hadacek)

for the benefit of potential clients. As a result, the AMX30 Bitube DCA was successfully offered by GIAT on the export market in 1975 and was ordered by Saudi Arabia as the AMX30SA. In Saudi service its job was to defend Shahine and Roland batteries from low-level attack by aircraft and helicopters. Some 53 vehicles were

AMX30-based engineers' vehicles

EBG (*Engin Blindé du Génie*)

The EBG is a 40-tonne engineers' vehicle based loosely on the AMX30D chassis, ordered in 1987. It was the only derivative of the AMX30 developed wholly by ARE without extensive reliance on other facilities. It was equipped with a dozer blade and hydraulic arm designed to mount a pincer, an auger or a rotary saw blade to accomplish battlefield engineering tasks. It was crewed by three men and was armed with

RIGHT The boring drill attachment could be used to install mines or to bury demolition charges in roadways. *(STAT)*

RIGHT The EBG's two-man turret mounted a 142mm demolition charge launcher. It was also armed with a 7.62mm machine gun mounted in a ball mount, which could be used to lay down suppressive fire. We can see how neatly the hydraulic arm folded alongside the right side of the hull, even with the pincer fitted. *(STAT)*

RIGHT The EBG prototype undergoing trials in snowy conditions. *(Jerome Hadacek)*

LEFT The hydraulic arm could be controlled from within the vehicle, or as seen here, remotely operated via an extension lead from outside the vehicle. (Jerome Hadacek)

LEFT The sharing of the AMX30B2 chassis with armoured regiments was intended to simplify the job of maintaining heavy AFVs at divisional level. The EBG was, as a consequence, an easy vehicle to drive and proved extremely versatile in service. (STAT)

LEFT The EBG's turret, seen here dismounted, was fitted with a 142mm demolition charge launcher, four mine-scattering tubes and a 7.62mm NF1 machine gun. The 142mm launcher was a single-shot weapon loaded from the muzzle, by crewmen outside the turret. (Jerome Hadacek)

ABOVE The relatively small series of 71 production EBGs were delivered over a four-year period between 1989 and 1994. The EBG's first use in action came in 1991 in support of the *6e Division Légère Blindée* in Operation Daguet. *(Thomas Seignon)*

BELOW A significant development of the EBG came in 2006, when a fleet of 42 EBGs were comprehensively rebuilt with Mack E9 diesel engines by Nexter. Twelve EBGs were stripped of their hydraulic arms and were fitted with an adaptation of the Israeli CARPET mine-clearing device. Known in French service as the SDPMAC, these EBGs carried a multiple rocket launcher box hinged at the rear of the chassis. *(Jerome Hadacek)*

a cupola-mounted 7.62mm machine gun and a fixed-traverse 142mm demolition gun.

Seventy-one EBGs were built at ARE on new AMX30B2 chassis from 1989 to 1994. The EBG was the only AMX30 variant designed from its inception with an automatic transmission. The EBG was expected to be replaced by the Leclerc-based EPG, but this vehicle was cancelled prior to production. As a consequence, in 2005 an order was placed to modernise 54 EBGs with the Mack E9 engine, 12 of which were ordered to be converted

into SDPMAC (*Système de Déminage Pyrotechnique pour Mines Antichar*) standard, which required substantial modification in order to mount the Israeli CARPET rocket mine-clearing system. The SDPMAC system fired 20 mine-clearing charges, which cleared mined areas through sympathetic detonation. The hydraulic arm was removed on these vehicles, with the crew reduced to two men as a result. Both types remain in service to the present day.

AMX30H *Poseur de Pont* bridge-layer

The first engineers' vehicle planned on the basis of the AMX30 chassis was the bridge-layer variant, designated *Char Moyen Poseur de Pont Mle F1*. This capable vehicle could quickly deploy a scissor bridge by means of a hydraulic ram and chain drives, crewed by three men (vehicle commander, driver and *pontonnier*). The hydraulics and bridge control apparatus were housed in a large casemate manufactured at the *Atelier de Tarbes* (which was crewed by the vehicle commander and the *pontonnier* (or bridge operator) located in the area occupied by the fighting compartment in the AMX30B. The bridge-layer design was completed in 1967 but series production was repeatedly delayed, before official adoption by the French Army in 1974 and then cancellation in 1975.

The bridge itself was designated *Travure Mle F1* and was manufactured by Coder. It weighed

8,600kg, was 22m long deployed and 3.95m wide. The 12 vehicles originally envisioned for the French Army were subsequently included in the sale of AMX30S battle tanks and variants to Saudi Arabia. A single vehicle was retained in France as a GIAT reference vehicle and was preserved in the Musée des Blindées de Saumur.

Char de Déminage Téléguidé

Faced with the French Army's shortage of armoured mine-clearing systems at the outset of Operation Daguet, GIAT equipped

ABOVE Because the standard EBG was still deemed essential in engineer units, the larger portion of the EBG fleet modified in 2004 retained the standard hydraulic working arm. Appliqué armour was fitted to both versions. The two EBG variants served alongside each other.
(Jerome Hadacek)

LEFT The SDPMAC fires its rockets forwards over the length of the vehicle, destroying minefields by sympathetic detonation.
(Jerome Hadacek)

five AMX30B battle tanks with ex-East German KMT5 mine-rollers (donated by the *Bundeswehr*) and remote driving controls for use in Kuwait. These conversions were designated AMX30EBD (*Engin Blindé de Déminage*), serving effectively in the liberation of Kuwait. These tanks retained their full armament and cleared minefields with the turret traversed aft. The rationale behind this type of vehicle was for the crew to operate the vehicle until a minefield was encountered, then dismount and employ the mine-clearing vehicle by means of remote control. The experience of employing the AMX30B as a remote-controlled mine-clearing vehicle during Operation Daguet led to the conversion of five purpose-built remote-controlled mine-clearing vehicles in the 1990s, designated *Char de Déminage Téléguidé* or AMX30B2DT.

ABOVE The AMX30H bridge-layer was designed to transport, launch and recover a 22m-long Class 50 scissor bridge. Although the French went to considerable lengths to procure the AMX30 *Poseur de Pont F1*, and went so far as to standardise and order a production series in 1974, these never saw service within the French Army. Facing budget shortfalls, the order was cancelled and instead they made do with the much smaller AMX13-based version purchased in 1967. The 12 production AMX30H *Poseur de Pont* vehicles were purchased instead by the Royal Saudi Army. *(Jerome Hadacek)*

BELOW LEFT Unlike other contemporary bridge-layers, the AMX30H laid its bridge from the rear of the chassis, under the control of the *pontonnier*. *(J.M. Boniface)*

BELOW RIGHT This method of bridge-laying was already well established on the much lighter AMX13 *Poseur de Pont F1* and was effectively scaled up. *(J.M. Boniface)*

Mine-clearing vehicles

Five surplus AMX30B2 chassis were converted to this purpose, fitted with surplus AMX30B turrets with blanked-off mantlets. The vehicles were equipped with Israeli Ramta mine ploughs and with URDAN mine-roller systems mounted on the front of the vehicle. In 1997 an external video camera system was added to increase the effectiveness of remotely clearing minefields. A second series of ten AMX30B2DTs were converted subsequently, all eventually equipped with the GIAT BS G2 reactive armour system in

LEFT One of the five EBD vehicles converted from AMX30Bs for use in Operation Daguet in 1991. The EBD was remote controlled (in direct line of sight only) and was equipped with an ex-NVA (East German Army) Russian-designed KMT5 mine-roller system. The rear of the hull mounted five red lamps to permit the crew to maintain visual contact in dusty conditions. *(Jerome Hadacek)*

LEFT The first AMX30B2DT conversion series of five vehicles were based on the lessons learnt with the EBD. The basic vehicle was lashed up from a disarmed AMX30B turret mount based on an AMX30B2 hull. The mine ploughs were ordered from the Israeli Ramta company. *(STAT)*

LEFT The early-pattern AMX30B2DT was supplemented by an improved version which included a far more advanced video camera system that allowed the operator to better see the tank's surroundings when clearing mines. *(STAT)*

RIGHT Demining trials with live mines were conducted by the STAT. This is an AMX30B2DT from the second production series of ten vehicles. It carried the same GIAT BS G2 explosive reactive armour system as the AMX30B2 Brennus. Israeli-designed URDAN rollers can also be used in place of the plough. Note that in this configuration no lane-marking system is fitted. *(STAT)*

LEFT The tank is remote controlled using a more advanced indirect line of sight device with direct video link permitting operation from 3,000m. *(STAT)*

BELOW LEFT Under a dusting of snow we can see the late-pattern AMX30B2DT without any mine-clearing or lane-marking equipment fitted. The TOP7 cupola mounts a mast which controls two video cameras. Additional cameras can be fitted to the turret sides. *(Thomas Seignon)*

BELOW RIGHT The AMX30B2DTs were operated by the *13e Régiment de Genie*. The vehicles retained deep-water-fording capability and the battery cover plate stowage was resited to the rear side of the turret. *(Thomas Seignon)*

RIGHT This heavy-lift Pinguely crane mounted on a specially modified AMX30B2 or EBG chassis was a unique prototype that never went beyond this stage. Note the reinforced wheel pattern and the exhaust extensions. *(STAT)*

order to protect the mine-clearing system from RPG attack in transit to the mined area – or to improve survivability during clearance operations. The AMX30B2DT could be remotely controlled by the crew from a VAB armoured personnel carrier to afford the crew some protection from enemy fire as well as from mine blasts. In 2014 some nine vehicles were still in service with the *13e Régiment du Genie*, but it is unknown how many are still operational at the time of writing.

RIGHT The ENFRAC (*Engin de Franchissement* – or crossing vehicle) was a unique prototype designed to facilitate river-crossing operations, able to prepare riverbanks, and – if necessary – bridge the gaps using its waterproof chassis. Presented to the army for trials in 1970, the concept never went beyond the prototype stage. *(Fonds Claude Dubarry, Collection du Musée des Blindées de Saumur)*

Chapter Six

AMX30 exports

France was denied its ambition to arm its FINABEL allies with the AMX30 by the success of the Leopard in the 1960s. After GIAT changed its strategy to market the AMX30 to Spain, Greece, South America and the Middle East, it quickly achieved substantial sales. Ultimately nearly 750 AMX30B and AMX30S were exported, 280 were license-built in Spain and over 500 AMX30 variants were also sold in Europe, the Middle East and South America.

OPPOSITE **An AMX30S on a proving test prior to delivery.**
(Fonds Claude Dubarry, Collection du Musée des Blindées de Saumur)

The AMX30 family in foreign service

Belgium and the Netherlands

France had already known the success of the AMX13 by 1963, and DEFA had very high hopes in its successor. By 1964 Belgium and the Netherlands had already done business with DEFA and SOFMA (DEFA's export arm). Both countries had bought significant numbers of AMX13-type vehicles, and indeed Belgium was manufacturing the VCI armoured personnel carrier under licence. The Netherlands Army (*Koninklijke Landmacht* – KL) endured a well-publicised technical scandal with the AMX13 in 1964. Substantial quality problems were found with their newly delivered AMX13 fleet which DEFA had to rectify. While Joseph Molinié and the French Army minister were heavily involved in solving the problems, this likely affected the KL's position on further acquisition of French-built AFVs. The KL already knew the 105mm Obus-G from their AMX13 Mle 58s and had also recently adopted the 105mm L7 gun for their Centurion Mk 5 and Mk 7s. West German military knowledge was highly valued in some quarters of the KL's officer corps, and West German evaluations of French equipment (in some cases negative, especially their conclusions regarding the AMX13 VTT) were shared with the Dutch. British influence on the KL's cavalry corps in the 1950s had also been substantial, and though the Dutch wanted a lighter tank than the Centurion, they had great respect for its gun. From these recent experiences and with the knowledge that four production lines had been established in West Germany, senior Dutch officers followed the Leopard's development with interest.

Belgium and the Netherlands had both been invited to observe the Mailly trials in October 1963. Belgium held the keener interest in the AMX30B for its army, and negotiations between the Belgian defence minister Poswick and the French Army minister Messmer were pursued from December 1963 to July 1966. Exactly how the prospect of a Belgian order for the AMX30B went astray is not recorded, but later that year the discussions ended and the Belgians were the first NATO ally to order the Leopard.

Unsuccessful deals

On 1 July 1969 the French government proposed an offer to the Italian government to co-produce the AMX30, which was also never concluded. In April 1970 the Danish defence ministry held discussions with the French government that included the AMX30B, but these too were unsuccessful. Both Italy and Denmark eventually purchased Leopards in the 1970s. The AMX30B was proposed unofficially to representatives of the Canadian, Australian, Argentinian and Israeli governments without securing further interest.

Spain

The only foreign production licence granted by GIAT for the AMX30 was to Spain. A presentation of the AMX30B was made in Paris by DTAT to representatives of the Spanish defence ministry in December 1969. The AMX30 was a second choice for the Spanish – ordered after efforts to buy the Leopard were blocked by British refusal to allow sale of L7A3 guns to the Franco regime. An agreement of military cooperation was signed shortly afterwards in June 1970 and the first 19 AMX30B tanks arrived in Spain in subsequent weeks. Also in June 1970, the first Spanish foreign legionaries left for France to receive instruction with this new materiel. One tank was kept in Madrid for training and tests.

The other AMX30Bs were transported to North Africa to form the *Bakali* medium tank company of the Spanish Legion, to serve in its deployment to the Spanish Sahara. The first six tanks arrived in Irun in November 1970, then moved by rail to Bilbao where they were embarked in the transport *Almirante Lobo*. They landed at Cabeza de Playa and drove to the base area at Sidi Buya in Laayoune, where they were repainted in sand-coloured paint. The 18 tanks of the *Bakali* company formed three sections of five tanks, with a headquarters equipped with three more. The *Bakali* company functioned as the tank company of the 3rd *Tercio* of the legion. When the Western Sahara conflict began in 1974, the *Bakali* company were joined by two companies of M48A1s from the 2nd Battalion, 61st *Alcázar de Toledo* Regiment as the armoured reserve of the Governor General of the Sahara. These French-built AMX30Bs proved reliable in the desert conditions.

After the Sahara crisis this force was withdrawn in December 1975. The *Bakali* company was dissolved and its AMX30Bs were transferred to the 55th *Wad-Ras* Mechanised Infantry Regiment. On 8 October 1971 a contract was negotiated for the co-production of the AMX30B in Spain, which included a production licence. An addendum was signed the following March covering further development of the design in Spain. The original 19 AMX30Bs were later modified to AMX30E standard.

The Spanish produced 280 AMX30Es, with deliveries from 1978 and ending during the course of 1983. During production the AMX30Es were equipped with the heavier torsion bar type developed for the French

ABOVE The AMX30E was very similar to the AMX30B, with detail changes to suit Spanish requirements. As production continued a higher proportion of the tank was manufactured in Spain. The engine, gun and turret casting were all manufactured in France. This is an AMX30ER1, remanufactured with the CD-850 transmission. *(Luis Pitarch Carrion)*

AMX30 derivatives and later adopted for the AMX30B2. The commander's machine-gun mount on the TOP7 cupola was modified to mount the Spanish Army standard MG42/58 (and later the MG3) machine gun. AMX30Es carried a distinctive antenna tube over the spare track link stowage on the right side of the hull above the pry bar used to test the suspension's torsion bars. Another difference from the AMX30B was the more pronounced bulge in the driver's hatch, which permitted slightly more head room for taller drivers. American AN/VRC radios were fitted as standard.

The Spanish Army's experience of operating the AMX30E did not closely mirror the positive results experienced by the French. Both armies had operated the M47 in the previous decade, though the Spanish were much less generously equipped in terms of maintenance facilities than the French. AMX30E production vehicles had a long history of mechanical unreliability and as a result they became unpopular with their crews. For example, the 14th *Villaviciosa* Cavalry Regiment, which normally left on manoeuvres with 21 tanks regularly recorded up to 5 tanks (roughly 25%) temporarily lost to transmission failures. The AMX30Es built in Spain experienced many

gearbox. Spanish sources also criticised the French tracks for their life of 3,000km. The first change attempted in Spain was to modify one of the French-built AMX30Bs to mount the AVDS 190 engine and Allison transmission mounted in the ongoing M48A5 programme envisioned for Spain's M48 fleet. By 1979, the Spanish Army had already set up a work group at ENOSA (Empresa Nacional de Óptica SA) to devise improvements to the AMX30E, a programme that mirrored efforts in France to adopt features from the AMX32. The ENOSA project evolved over the space of several years, proposing at first to simply install the same Minerva ENC200 automatic gearbox newly developed for the AMX32 (and adopted for the AMX30B2). This option was studied with trials of modified vehicles in three of Spain's armoured regiments, but it was not pursued to a production series. One of the options examined by the ENOSA team to modernise the AMX30E's powertrain in the wake of the 1979 directive had included a 720hp MTU MB833 Ka 500 diesel with a ZF MP250 gearbox, an 800hp GM124-71 diesel with Allison CD-850-6B automatic gearbox, the AVDS-1790-2C diesel and CD-850-A gearbox – similar to the arrangement employed in the US M60A1 – and a 720hp HS-110 matched to a Renk automatic gearbox. In all, five motor/

ABOVE The first Spanish experiment in adjusting the AMX30's powertrain was to modify one of the AMX30Bs returned from the *Bakali* company. This was modified by Talbot to mount an AVDS 1790 2A engine and CD-850-6A transmission as used in the modification of the Spanish M48A1s and M47s (and was nicknamed *El Niño*). *(Pedro Miguel Paniagua)*

problems with the transmission and clutch, which worried the army high command – and solutions were sought.

The AMX30E suffered frequent problems with the HS-110 engine and the BV-5-SD manual

RIGHT Modifications were only made to the engine compartment, which required lengthening the hull and consequently longer tracks. The engine decks and rear hull plate were extensively modified. New air filters compatible with the American diesel were also fitted. It may have served in some measure as inspiration for the Venezuelan Army's later project to modernise their AMX30B's powertrain. *(Pedro Miguel Paniagua)*

gearbox combinations were tested out before the Spanish Army arrived at a satisfactory and affordable combination of the existing HS-110 engine and an American Allison CD-850-6A automatic gearbox. The new French 105mm OFL-F1 kinetic energy round was also adopted in 1982.

In the meantime, equipping of units with the standard production AMX30E continued through 1978–79; in the Spanish 1st *Brunete* Armoured Division the 61st *Alcázar de Toledo* replaced their M48 and M48A1s with AMX30Es. The 55th *Wad-Ras* Mechanised Infantry Regiment received its first AMX30Es a few months later, completing its establishment with the French-built ex-*Bakali* company tanks. The 16th *Castilla* Mechanised Infantry and 14th *Villaviciosa* Cavalry Regiment were also re-equipped with AMX30Es in due course. In the following years the 3rd Maestrazgo Division's 21st *Vizcaya* Mechanised Infantry Regiment (in Betera and Valencia), and the 18th *España* Mechanised Infantry Regiment (in Cartagena and Murcia) were also equipped with companies of AMX30Es. Additional vehicles were provided to the infantry and cavalry academies for the training of future officers and non-commissioned officers.

With 299 AMX30Es in service by 1984

ABOVE The first series conversion ordered on the Spanish AMX30E park involved the replacement of the original transmission, a conversion designated AMX30ER1. In order to accept the American CD-850-6A transmission, the engine compartment plates had to be extended upwards in order to accommodate a taller powerpack unit. In addition, a slave cable socket (electrical outlet for external start) was fixed into the left-hand headlamp group. A new jerrycan stowage pattern was adopted for the rear plate, with three jerrycans carried vertically. The fire controls remained identical to those of the AMX30E. *(Luis Pitarch Carrion)*

BELOW The AMX30EM2 (1987) was a comprehensive modernisation of the AMX30E, addressing the tank's fire controls and powertrain. *(Pedro Miguel Paniagua)*

and the recurring problem of vehicle availability from transmission failure, the decision was taken to upgrade 60 of the AMX30Es with the well-regarded American Allison CD-850-6A transmission as AMX30ER1s. An identical upgrade was applied to the Spanish Army's AMX30Ds first delivered in 1978. In 1987 the MTU MB833 Ka 500 diesel with a new ZF LSG 3000 gearbox from West Germany (as employed in the Marder IFV) was selected to upgrade 150 more AMX30Es into AMX30EM2s. The AMX30EM2 also incorporated the suspension upgrades specified for the French AMX30B2 conversions. A modernisation of the turret systems was also provided for in the AMX30EM2, producing a substantially more capable weapon system than the production AMX30E.

BELOW An AMX30EM2 on manoeuvres. With the new powerpack and its relatively light weight, the AMX30EM2 enjoyed an excellent power-to-weight ratio and cross-country performance. The AMX30EM2s served in the 1st *Brunete* Armoured Division, 12th BRIAC and 9th BRIMZ, as well as in the 11th *Espana*, the 4th *Pavia* and the 9th *Numancia* Cavalry Regiments. Smaller numbers served in the cavalry academy and briefly in the regiments of the 1st *Jarama* Cavalry Brigade (where they were quickly replaced with the VRC105 Centauro armoured reconnaissance vehicle). It was planned to modify the entire fleet of over 290 vehicles to AMX30EM2 standard, but the arrival of surplus US M60A3TTS, as well as the Treaty on Conventional Armed Forces in Europe of 1990, resulted in the upgrade programme being cut to 149 vehicles. (Jesús Pardo)

RIGHT This AMX30EM2 was registered as 'ET VE 98426', and was prepared for mounting the SABBLIR armour system, which was never fitted. Among the other modifications that came with the provisions for the SABBLIR system were the installation of larger stowage frames on each side of the turret. Unlike the AMX30B2 Brennus developed in France, the AMX30EM2 SABBLIR included ERA panels on the hull side armour.
(Angel Ruiz)

Improvements included the adoption of the American Hughes Mk 9A/D fire control system built under licence in Spain by ENOSA (including a laser rangefinder and ballistic computer inspired by the system adopted for the M1 Abrams) with new commander's and gunner's sights. All 149 AMX30EM2s were equipped with thermal sights, which changed the layout of the optics on the turret roof, while a barrel colimator and wind sensor were also installed. Four 76mm Wegmann smoke dischargers were added to each side of the turret, dust skirts were installed and the loader's hatch was modified to mount an AMP M2 HB 12.7mm machine gun.

Much as had already been envisioned in France, ENOSA also proposed to improve the AMX30EM2's protection with explosive reactive armour (ERA). The first type tested at Cádiz on two AMX30EM2s in 1990 was designated SBBR (*Santa Barbara Blindajo Reactivo*). An improved type designated SABBLIR (the acronym changing subtly to *SAnta Barbara BLIndajo Reactivo*) was perfected and under test by November 1991. The implementation of the Spanish Army's *Plan Coraza* (which included the purchase of the Leopard 2A4 and the production of the Leopard 2E) also rendered the requirement for the upgrade of AMX30EM2

RIGHT Registered as 'ET VE 97963', this vehicle was the only AMX30EM2 demonstrator completed for the SABBLIR armour system. Like the French GIAT BR G2, SABBLIR was a lightweight reactive armour system. In the case of the SABBLIR system, ceramic plates were incorporated in each brick, combining lightness with excellent anti-HEAT characteristics. *(Angel Ruiz)*

ABOVE The SABBLIR explosive reactive armour kits proven on the AMX30EM2 were further developed for use on the Centauro armoured car, which replaced the AMX30 in Spanish service.
(Angel Ruiz)

RIGHT The SABBLIR installation on the forward slope of the turret roof and on the gun mantlet.
(Angel Ruiz)

with the SABBLIR system redundant. The AMX30 thus passed from Spanish service. Negotiations to sell the surplus AMX30EM2s to Indonesia and Colombia were not successfully concluded and the bulk of these vehicles were also scrapped as the Leopard 2s were delivered in the 1990s.

Greece and Cyprus

The second order from Europe was secured in May 1970 from Greece, which ordered 190 AMX30Bs and 14 AMX30Ds, with deliveries commencing in 1974. The Hellenic Army's AMX30s were very similar to the original production French Army AMX30Bs, mounting the 12.7mm coaxial machine gun and equipped for submerged river crossings, but all were delivered equipped with sand skirts. They received very few modifications in a service life of 20 years, some 40 being transferred to the Cypriot National Guard after they were withdrawn from Hellenic service. Cyprus was the single verified customer to buy the AMX30B2, with 52 tanks from the 8th

ABOVE The Hellenic Army adopted a number of disruptive camouflage schemes, like the one worn by this AMX30B in 1985. *(T. Metsovitis)*

BELOW An AMX30B being demonstrated to the Hellenic Army in 1970. The Greek government ordered 190 AMX30Bs, as well as 14 AMX30Ds, with deliveries from 1974. The AMX10P infantry combat vehicle was also ordered, allowing a complete Greek armoured brigade to be equipped with French-produced equipment. *(Bernard Canonne)*

LEFT Following the end of the Cold War, at least 40 AMX30Bs were transferred to the Cypriot National Guard by the Hellenic Army. *(Stelios Markides)*

ABOVE The Cypriot National Guard was the only export customer for the AMX30B2. *(Proelasi)*

BELOW The AMX30V was unique in that it featured a gun stabilisation system as part of its fire control system upgrades installed by SABCA. *(Carlos Antonio Arroyo Alonso)*

conversion batch delivered in 1990–91. These vehicles were decommissioned in 2018.

Venezuela

In South America, Venezuela ordered 81 AMX30Bs and two AMX30D recovery vehicles in 1971, with the first deliveries within the year (other sources speak of 86 AMX30B, 2 AMX30D and 4 AMX30H deliveries reported in SIPRI documentation, but delivery is unverified – and the bridge-layers very likely refer to the earlier AMX13 *Poseur de Pont F1*). The Venezuelan AMX30Bs were extensively upgraded in 1985 into AMX30Vs. The conversion replaced the M208 and M270 optics with a Belgian SABCA fire control system (including new sights, a laser rangefinder and full gun stabilisation). The original HS-110 engine and transmission were replaced by the AVDS1790 diesel and Allison automatic transmission as fitted to the M60A1, and enlarged fuel tanks were also installed. This required a lengthening of the rear hull, visible between the No 5 road wheel and the final drives, much like the *El Niño* prototype attempted in Spain.

Unique among AMX30 upgrades, the AMX30V also included the Model 2880 hydropneumatic suspension system. This replaced the original torsion bars and reportedly offered the rebuilt vehicles a far better cross-country performance. The AMX30V retained the original single-pin tracks. As such, the AMX30V compared very favourably with the later AMX30B2 conversions or with the AMX30EM2, all deficient only in their lack of modern armour protection. The AMX30V could have been easily improved further with the GIAT or SABBLIR explosive reactive armour at a small weight penalty (although this path was not followed by the Venezuelan government). The AMX30V is still believed to be in service with the Venezuelan Army, deployed in proximity to the Colombian border in the 11th Armoured Brigade. It is believed that several

RIGHT The AMX30V's suspension was upgraded with a hydropneumatic system, while the powertrain was essentially a modified version of that employed in the US M60 series tanks. The lengthened hull was similar to Spanish modifications tested out in 1978. *(Carlos Antonio Arroyo Alonso)*

large deliveries of AMX30 spares have been negotiated with Spain and France since 2006.

Chile

Chile was another country deeply interested in the AMX30B in 1970, ordering 50 before the Pinochet coup in 1973. The French government refused to complete the order for the new Junta, and then met considerable criticism when it finally honoured part of Chile's order, delivering 21 tanks in 1981. These appear to have been shipped under cover of darkness from Bassens on the Liberian-flagged *Lif-Ton*, a fact deplored in the *Assemblée Nationale* on 20 April 1981. These were probably among the last AMX30Bs built at ARE, alongside 60 for the French Army still on order in October 1981. In 1998 at least 21 more surplus AMX30Bs and 10 AMX30Ds were sold to Chile from French Army stocks (along with a substantial stock of spares), which completed the original Chilean order in the broader sense.

Saudi Arabia

The AMX30 was adopted by the Royal Saudi Army, which began negotiations to buy a modified AMX30 (known as the AMX30S) along with a range of other weapon systems in April 1970. This resulted in a long and fruitful partnership between GIAT and Saudi Arabia, and the sale of 290 AMX30S, 58 AMX30D, 12 AMX30H *Poseur de Pont* bridge-layers, 52 AMX30 Bitube 30mm DCA and 51 Au 155 GCTs. They also ordered the AMX30R Roland 2 and the Shahine missile systems. Deliveries began in 1972 and continued

ABOVE The Royal Saudi Arabian Army was the largest foreign customer for French-manufactured AMX30s, and purchased a dedicated version known as the AMX30S. These vehicles were part of a larger arms deal that extended to missiles and many other types of military vehicles. This AMX30B was employed in trials in Saudi Arabia to determine the necessary modifications. *(Fonds Claude Dubarry, Collection du Musée des Blindées de Saumur)*

RIGHT The AMX30B tested in Saudi Arabia included a number of modifications to facilitate use in the desert, most obviously the use of extended trackguards. *(Fonds Claude Dubarry, Collection du Musée des Blindées de Saumur)*

LEFT The 105mm gun and the 12.7mm coaxial weapon are both fitted with muzzle covers to minimise the ingress of dust in sandy conditions. *(Fonds Claude Dubarry, Collection du Musée des Blindées de Saumur)*

CENTRE Fresh from the ARE assembly line in France, an AMX30S is put through its paces. *(Fonds Claude Dubarry, Collection du Musée des Blindées de Saumur)*

BOTTOM Note the AMX30S's extended track guards and the sheet metal sand shields fitted below the turret ring, which were intended to minimise the amount of dust sucked into the engine's air intakes. *(Fonds Claude Dubarry, Collection du Musée des Blindées de Saumur)*

into the early 1990s, and these sales were especially important to GIAT for they included long-term service contracts. The Royal Saudi Army employed the AMX30S and its variants during the liberation of Kuwait, and has since reportedly employed the AMX30S in action in Yemen. Accurate information regarding Saudi military matters is scarce – particularly concerning the Saudis' use of their AMX30-based artillery and air defence variants. The current status of the Saudi Au 155 GCT and their once-large force of AMX30-based self-propelled anti-aircraft systems is unknown.

United Arab Emirates

The United Arab Emirates ordered 64 AMX30Bs in 1972 with four AMX30Ds, which were all delivered within a two-year period. Following the adoption of the Leclerc EAU in the 1990s and the UAE's participation in multiple UN peacekeeping missions, some 50 Emirati AMX30Bs were gifted to Bosnia. It is unknown if any remain in service.

Qatar

In 1977 the Qatari government purchased 24 AMX30s and a single AMX30D. The Qataris deployed the AMX30B in the defence of Saudi Arabia and the liberation of Kuwait in 1991. The small Qatari tank battalion, made up of only two companies of AMX30Bs, saw battle in the defence and subsequent recapture of the Saudi

city of Al Khafji. The Qatari tank battalion, some 24 tanks strong, first fought in support of Saudi National Guard infantry on 30 January 1991. The USMC's 2nd Light Armored Infantry Battalion and a small task force of tanks from the 2nd Armored Division's Tiger Brigade reinforced the Saudis. The AMX30Bs were deployed to stem the Iraqi attack south of Al Khafji, and they recorded at least five Iraqi T-55s destroyed that day. One of the Qatari AMX30Bs was also destroyed in battle on the same day, with the loss of its entire crew. Over the following two days the Qatari tanks fought alongside US Marine and Saudi forces to liberate the city, and suffered a second AMX30B destroyed. Ultimately airpower decided the battle at Al Khafji, like elsewhere in the first Gulf War.

The single production line at ARE was kept busy throughout the 1970s filling French and foreign orders before switching to the production of new AMX30B2s between 1982 and 1986. Thereafter, the production of long-delayed support and artillery versions at ARE (as well as the reconstruction of AMX30B2s) maintained a full order book until the end of the Cold War at Roanne. A very large pool of redundant AMX30Bs and AMX30B2s was assembled to serve as a source of spares. Some of the best 'used' AMX30Bs were rebuilt after the Cold War by GIAT Industries and sold to Qatar and Chile in 1998.

ABOVE A larger set of extended track guards was also tested, but was not retained for the AMX30S. *(Fonds Claude Dubarry, Collection du Musée des Blindées de Saumur)*

LEFT Here we can see a colour photo of an AMX30S being tested in the desert in Saudi Arabia. The amount of dust raised by a tracked vehicle moving at speed in such conditions is evident. The AMX30S entered service in 1975 and has served for decades. These tanks were employed in action in the liberation of Kuwait in 1991 and have seen service as recently as 2019 in the conflict in Yemen. *(Fonds Claude Dubarry, Collection du Musée des Blindées de Saumur)*

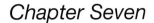

Chapter Seven

The AMX30 series in combat

In 1991 the AMX30 and AMX30B2 were blooded when the Saudi, Qatari and French armies were deployed to evict the Iraqi Army from occupied Kuwait. In French service, crewed by professional soldiers, the AMX30B2 had little problem dealing with the Iraqi T-55, T-59 and T-62s they encountered in the desert wastes. Qatari and Saudi AMX30s also saw heavy combat around Al Khafji.

OPPOSITE The AMX30B2 saw its first and only combat with the French Army during the 1991 Gulf War. *(Thomas Seignon)*

Service in the French Army

The AMX30B never saw combat in the French Army, but its successor the AMX30B2 earned its spurs honourably in battle against Iraqi forces during Operation Daguet in early 1991. The AMX30B2's first experience of combat came at the end of the Cold War at the spearhead of the *6e Division Légère Blindée* (6DLB). The 6DLB was raised strictly from units already incorporating professional soldiers and was made up to war establishment with conscript volunteers from every arm of service. In 1991 the French Army was composed of young men called up for 12 months of obligatory military service. A few units were partly or completely made up of professional soldiers, most of which were assigned to the *9e Division d'Infanterie de Marine* or to the *6e Division Légère Blindée*, formations normally assigned to foreign deployments.

The French government had drawn regularly from both of these formations for units deployed in Africa, and many of their regiments included combat veterans. With the exception of a single squadron of the 501e RCC deployed for manoeuvres in Senegal in 1980, the AMX30B had never served in French colours outside France or West Germany. The French Army's foreign deployments in the 1970s and 1980s had long relied on the

little Panhard AML for armoured firepower. The difference for Operation Daguet was that the Iraqi Army in 1991 included a substantial armoured corps equipped with T-55, T-62 and T-72 MBTs. The *6e Division Légère Blindée* was normally a wheeled armour formation, featuring units equipped with the AMX10RC reconnaissance vehicle as its most potent AFV. During the Kuwaiti crisis of late 1990 the French government adopted the army's recommendation to assign an armoured regiment equipped with tanks to the division at a very late stage.

Facing a potential enemy using modern Soviet battle tanks, the decision to incorporate one armoured regiment raised from the two professional heavy armour squadrons available to the *Arme Blindée Cavalerie* made a great deal of sense. The two professional squadrons formed part of two regiments: the *4e Régiment de Dragons* and the *501e Régiment de Chars de Combat* (these squadrons were known by their abbreviated titles of 1/4 RD and 2/501 RCC). The latter of these two units was actually equipped with AMX30Bs and its personnel had to be rushed through conversion courses in order to participate. This marriage of circumstance was known collectively as the *4e Régiment de Dragons* (4e RD) throughout the operation (and it remained constituted as a unit thereafter, the rest of the 501e RCC being dissolved in the army reorganisation that followed the end of the Cold War).

The 4e RD was equipped for deployment with 44 of the most modern, late-production AMX30B2s in the *Arme Blindée Cavalerie*. The tanks selected came from the collective inventory of regiments recently re-equipped with the AMX30B2 model in order to ensure that all were using the DIVT16 thermal camera. Additional equipment specific to the operation was added during a quick overhaul after 44 tanks had been assembled at Mourmelon, not far from the city of Reims.

The 4e RD was organised into three armoured squadrons and a single command and logistics squadron (*escadron de commandement et logistique* or ECL; equivalent broadly to a British headquarters squadron). Each of the armoured squadrons included a command tank, three four-tank platoons

and a support platoon mounted in three VAB 4 × 4 armoured personnel carriers. The ECL incorporated all of the support platoons employed within the regiment, including logistics, medical, anti-aircraft, quartermasters and of course the regimental headquarters, which

ABOVE A crewman guides his driver as he backs off a tank transporter in Saudi Arabia during the preparations for the ground assault in early 1991. *(Thomas Seignon)*

BELOW A squadron of AMX30B2s in column in early 1991, probably prior to moving to an assembly area. *(Thomas Seignon)*

ABOVE Much of Operation Daguet was spent maintaining equipment, doing system checks and NBC drills prior to the ground phase of the war. The AMX30B2 performed extremely well in the desert and many of the new features added frantically prior to departure (like the LIR30 and the Galix smoke grenade launchers) saw relatively little use due to the rapid collapse of enemy morale. *(Thomas Seignon)*

BELOW The CN-105-F1 firing the OFL-105-F1 proved perfectly adequate to destroy most enemy MBTs it engaged, although no T-72s were encountered. *(Thomas Seignon)*

included the commanding officer's tank and a reserve platoon of four tanks. The organisational model chosen for the regiment was a modified version of the RC40 order of battle that was being evaluated, to represent half of the future 80-tank regimental organisation expected for the next-generation French MBT.

While the invasion of Kuwait by Iraqi forces in late 1990 is well documented elsewhere, the American-led coalition's operation to liberate the oil-rich country in early 1991 was conducted in two phases. The first of these was an air campaign that began on 17 January 1991. The objective was to neutralise Iraqi military capabilities in depth, and to destroy the Iraqi Army's infrastructure and command apparatus. The air campaign was aimed clearly at paving the way for a comprehensive ground assault, which comprised the second phase of the plan. This was launched at dawn on 24 February 1991, with the objective of destroying the Iraqi Army and driving them back into Iraq. In the second phase, the US XVIII Corps was tasked to advance along the western flank of the coalition force. The 6e DLB was included in the XVIII Corps and was known in France as *La Division Daguet* after Operation Daguet.

The 4e RD's tanks received new Galix smoke dischargers, and the LIR30 infrared lure system to jam anti-tank guided missiles. Command tanks were equipped with an early form of global positioning system. Sand skirts were added to cover the upper run of tracks in the hopes of minimising ingress, and the tanks were repainted in a sand and brown disruptive scheme prior to shipment from Mourmelon. The regiment departed from Mourmelon starting on 22 December 1990, and headed to Saudi Arabia. Their tanks were shipped by rail to the Mediterranean coast for loading into ships on 24 and 25 December 1990. The journey was a speedy one, passing through the Suez Canal on the night of 30/31 December and arriving in the

Saudi Arabian port of Yanbu on 2 January 1991. Here the tanks were offloaded and parked in an assembly area for transport to the front.

Royal Saudi Arabian Army tank transporters subsequently carried the AMX30B2s some 1,100km to the Miramar assembly area, located roughly 30km from the Iraq and Kuwaiti border areas. The division concentrated in the area from 11 to 17 January, before advancing to its deployment area, Olive, some 300km westwards, on 18 January. The division now sat at the very western edge of the coalition force and here it waited for a month for its orders to advance. The time was spent acclimatising the men and in preparing their officers for the battle that lay ahead.

The French division deployed in two battlegroups. The western group was headed by the *1e Régiment Etranger de Cavalerie* (1 REC) and the *1e Régiment de Spahis* (1 RS) mounted in AMX10 RCR armoured reconnaissance vehicles. The eastern group included the 4e RD's AMX30B2s and the VAB-mounted *3e Régiment d'Infanterie de Marine* (3 RIMa). The Iraqi 45th Infantry Division was expected to be the first enemy formation encountered. Its positions were held by two brigades spread across three fortified areas in depth, with a third brigade in reserve.

After an artillery barrage that covered the insertion of scout units and Special Forces well ahead of the main force, the division began its advance on 24 February 1991. The 4e RD advanced at 05.30hrs and first encountered

enemy forces some four hours later at 09.30hrs, in daylight. The AMX30B2s made good use of their 20mm CN-20-F2's secondary armament against enemy infantry and armoured personnel carriers, conserving their 105mm ammunition for any tanks which might appear. The advance was speeded by very effective artillery preparations which successfully neutralised many of the enemy positions. By mid-afternoon the regiment was over 20km inside the Iraqi positions. The crews made good use of the AMX30B2's mobility to rapidly outflank the static Iraqi infantry positions, but all observed strict orders not to fire their 105mm guns into Iraqi artillery positions for fear of detonating chemical rounds expected to be stockpiled in forward positions. Very few enemy tanks were

ABOVE The Thales DIVT16 proved an excellent piece of equipment which gave the AMX30B2 crews the capability of identifying enemy vehicles at long range in the dark or in bad weather. *(Thomas Seignon)*

BELOW Most of the AMX30B2s retained their PH8B mantlet searchlights for use at night, despite their age. *(Thomas Seignon)*

ABOVE Over the four days in action, the 4e RD never lost a single tank nor a single soldier. The 6e DLB's losses for the entire operation was two men killed in accidents. The AMX30B2 demonstrated 100% availability and excellent reliability in combat. The AMX30B2s had proved themselves decisive in action, served by some of the army's best trained crews. *(Thomas Seignon)*

encountered, and for the most part 105mm high-explosive rounds made up the regiment's ammunition expenditure.

On the morning of 25 February 1991 an Iraqi counter-attack consisting of a squadron of T-59 medium tanks was engaged and annihilated by the regiment's AMX30B2s as it approached the Iraqi 45th Division's main fortified area. Other

Iraqi tanks (largely dug into defensive lines) were engaged and destroyed throughout the day without loss to the 4e RD's squadrons. By the end of the second day the French division had penetrated 60km into the Iraqi positions, completely disrupting its defensive capability. The town of As Salman was surrounded on the third day by the eastern battle group, the 3 RIMa's infantry companies taking the town by assault while the 4e RD were tasked with fire support as needed, a mission that lasted approximately 24 hours. On the morning of 27 February the 4e RD prepared to resume its advance northwards but the news of a ceasefire arrived the following day at 06.15hrs.

RIGHT War over and suitably cleaned up, an old warrior rests. This tank, 648-0010, was built as an AMX30B in 1968 and was rebuilt into an AMX30B2 in the later conversion batches. After it returned from the deserts of Kuwait, it was repainted for the 1991 Bastille Day parade, serving until 1995, when it was rebuilt again as an AMX30B2 Brennus. *(Thomas Seignon)*

The regiment remained in proximity of As Salman until 22 March while awaiting transport back to France. The regiment arrived back in Mourmelon on 20 April 1991.

Elsewhere in the coalition, Qatari AMX30Bs had played a decisive part in the Battle of Khafji, and Saudi AMX30S and Au 155 GCTs had served throughout the Saudi deployment. Qatari losses amounted to two tanks, while

ABOVE LEFT Saudi soldiers during an NBC drill with their AMX30S on 19 January 1991 in Saudi Arabia, not far from the Kuwaiti border. The AMX30Ss were equipped with different sights and were delivered with or retrofitted with the CN-20-F2. Ten days after this image was taken the Qataris, Saudis and US Marines engaged Iraqi troops advancing into the Saudi city of Khafji. *(Georges Merillon/Gamma-Rapho via Getty Images)*

ABOVE The Battle of Khafji, lasting from 29 January to 1 February 1991, was the first major ground engagement of the Gulf War. The Iraqi advance was met by coalition ground forces with overwhelming air support. The Qatari AMX30s were built to a pattern very similar standard to the French AMX30Bs manufactured prior to 1974, but were delivered with sand shields. This was one of two Qatari AMX30s destroyed by Iraqi forces during the battle. The Qataris deployed their entire tank strength in a single small battalion of two tank companies. *(Patrick Durand/Sygma via Getty Images)*

BELOW A Saudi AMX30S seen in column with AMX10P infantry fighting vehicles in the vicinity of the city of Khafji after the battle. *(Patrick Durand/ Sygma via Getty Images)*

RIGHT The adoption of the CN-20-F2 on the AMX30S led to the installation of a blast shield around the base of the left side of the main armament, a feature never adopted by the French. *(Georges Merillon/ Gamma-Rapho via Getty Images)*

BELOW Saudi AMX30Ss and their crews during the 1991 Gulf War. The AMX30S remains in service with the Saudi National Guard to the present day. *(Georges Merillon/ Gamma-Rapho via Getty Images)*

Saudi losses, if any, remain unknown. The Iraqi Army's Au 155 GCTs sat immobilised on home soil for a lack of spares during Desert Storm, and were eventually neutralised by coalition air strikes.

In the liberation of Kuwait of early 1991 the AMX30B2 showed itself as a mature and effective design in combat. The *Arme Blindée Cavalerie* and the general staff knew, however, that the faithful 'X30' had reached the end of its development potential. Even though the agreed programme to re-engine the AMX30B2 with the Mack E9 engine and the programme to increase its armour through the Brennus programme remained to be accomplished, the AMX30's replacement was inevitable.

AMX30 ACRA
(Arme Anti-Char RApide)

In 1962 the *Etat-Major de l'Armée de Terre* (EMAT) issued a requirement for a combined gun/high-velocity missile launcher for use as a tank armament. The specification was complete by 1964 and development by the APX bureau proceeded quickly, with testing beginning in 1968. The importance of the ACRA as an armament system for the AMX30 was discussed at length in the national assembly in 1964 and for several years the ACRA was expected to be ordered as soon as development was completed. The APX gun/missile-launcher system developed for mounting in the specialised T142 cast turret for the AMX30 was known as the *Arme Anti-Char RApide* (which translates as a high-velocity anti-tank missile).

Between 1969 and 1971 the T142 turret and its fire controls suffered numerous delays and the ACRA lost much of its support in the *Arme Blindée Cavalerie* (ABC). In 1971, the army resolved not to procure the AMX30 ACRA, although the programme was still completed to a level where production could be pursued if necessary. The T142 turret for the AMX30 was completed, validated and was extensively tested out in 1973–74. By this time

ABOVE In the mid-1960s the AMX30 ACRA was expected to confer long-range killing power to the armoured regiments of the future. The T142 turret was served very much like the AMX30B, but its gunner served to aim and fire the missiles. The ACRA launcher system used a brass cartridge when firing the missile or as a conventional gun, ejected when the breech was opened. It could fire a low-velocity high-explosive 142mm round and the 140mm missile through its 4,200mm-long smoothbore barrel. The 142mm round weighed 20kg and was intended for the destruction of unarmoured targets. The gun/missile launcher's recoil travel differed depending on the type of ammunition fired: 240mm when firing the missile, 350mm when firing 142mm high-explosive *MUnitions Complémentaires* (MUCs). The 142mm barrel and chamber were rated for 3,000 bars of pressure when firing. The weapon could be depressed to -8 degrees and elevated to +20 degrees hydraulically. The missile could be fired to a minimum range of 100m and to a maximum range of 3,300m. By 1969 its cost and complexity had imposed delays and funding problems. Despite its cancellation in 1971, development of the T142 turret nonetheless continued until 1974. *(Fonds Claude Dubarry, Collection du Musée des Blindées de Saumur)*

the pressure to cancel the ACRA outright in favour of the simpler and less expensive *haut subsonique optiquement téléguidé tiré d'un tube* (HOT) missile could no longer be resisted. In 1975 the programme was cancelled and today the T142 turret survives in the Saumur collection. We can hope that the T142 turret will be displayed one day on a suitable AMX30B chassis, a memorial to the risks of high-technology weapon development.

Postscript: the GIAT AMX40

BELOW *The AMX40 prototypes were designed with considerable West German automotive content. It was GIAT's intent to offer the vehicle with the option of both a West German powertrain or with the French Poyaud V12X with the ESM500 transmission. (Nexter/Collection du Musée des Blindées de Saumur)*

The AMX32's failure to attract foreign orders following the French Army's decision to procure the AMX30B2 led GIAT's AMX-APX design team to consider a larger chassis as the basis for a 40-tonne tank armed with a 120mm gun in 1983. The AMX40, as the design was quickly baptised, existed alongside the AMX32 as a viable project that could be produced for the export market. The turret was based on that of the AMX32, but now incorporated a fully stabilised main armament and fire control system. The tank was designed from the outset to mount the CN-120-G1 120mm gun (with 40 rounds) and retained the CN-20-F2 secondary armament. It was intended for potential customers in the Middle East, specifically Saudi Arabia, Egypt, Iraq and the United Arab Emirates. Its competitors included the Vickers Valiant and Mk 7 MBTs, the Brazilian ENGESA Osorio and the Soviet T-72.

The AMX40 was expected to weigh 45 tonnes. The chassis was larger than the AMX30's, with six trailing torsion bars and six paired road wheels per side. The road wheels, idlers, sprockets and tracks were directly derived from those employed on the AMX30 series. It had an all-welded hull with a much larger engine compartment than previous designs, in anticipation of fitting a more powerful engine than previously mounted in any French tank design in order to achieve higher mobility. The hull was 6.8m long and 3.18m wide, smaller than contemporaries like the Leopard 2A4, Challenger 1 or M1 Abrams. The driver sat in the front left of the hull with stowage for 21 × 120mm rounds to his right. The Domange-Jarret company developed hydraulic rotary shock absorbers for the AMX40's suspension system to permit a high cross-country speed. The hull was fitted with composite armour on the glacis and could be fitted with armoured skirts over the front portion of the suspension.

The AMX40 turret's armour protection was heavier than that of the AMX32 too, with a composite armoured mantlet and spaced armour protecting the front parts of the turret. While the maximum thickness of the armour has not been revealed, the AMX40 was not so heavily armoured as to impede its high mobility, which remained a crucial asset in the overall French concept of a tank's protection. The stowed turret ammunition was equipped with a large blow-out panel in the roof. The turret sides were armoured to resist 23mm armour-piercing rounds. The COTAC fire control system employed in the AMX32 was paired with a fully stabilised main armament, gunner's sight and the M527 commander's panoramic sight.

The combined unit was referred to as the COSTAC (COnduite de tir STabilisée Automatisée pour Char). It retained the same APX-developed gunner's sights as the AMX32 (and carried by

LEFT **The AMX40's hull and fire controls benefited from much developmental sharing with the French ECP (*Engin Principal de Combat*) programme – the future Leclerc tank.** *(Jerome Hadacek/Collection du Musée des Blindées de Saumur)*

P2 and P3 were both fitted with structurally identical turrets, so identifying this prototype is difficult. The new Galix smoke grenade launchers could fire smoke grenades or anti-personnel grenades. *(Thomas Seignon)*

the AMX30B2, by then already in service). The DIVT13 low-light television camera system was replaced by the DIVT16 CASTOR, a thermal television camera displayed to the gunner's and commander's positions (and inherited by the AMX30B2 in due course). The commander's cupola was adopted along a similar pattern to that employed on the AMX32, with the ammunition box running around the lower edge with the commander's AN F1 machine gun mounted at cupola roof height to ensure that the M527 panoramic sight unit was not obscured.

The AMX40 carried 1,300 litres of fuel and enjoyed a range of 600km, hitting speeds in excess of 70km/h on metalled roads. Four hull prototypes were built, each differing in powertrain. The AMX-APX design team focused heavily on providing the vehicle with the means of accelerating very rapidly, conferring the AMX40 with excellent agility for a tank of its size. The first tests of the CH1 (Chassis 1) prototype hull in 1983 were accomplished with a boosted 850hp MTU 833 engine. The AMX30 series' single-pin 570mm-wide track was retained for the initial testing phase. The revised P2 prototype delivered in early 1984 (consisting of the CH2 hull prototype and the T2 turret) was fitted with a 1,100hp Poyaud V12X diesel equipped with a Renk ZF LSG 300 transmission (with four forward and two reverse speeds). The otherwise identical P3 prototype was fitted with an MTU 833 engine instead of the Poyaud. The first trials began in January 1985 at Valdahon, followed by firing trials at Mailly that May and December. One of the first changes made as a result was to alter the ammunition stowage to permit 12 of the 19 rounds stowed in the turret bustle to face nose-outward toward the breech. This permitted quicker handling and a higher rate of fire. A host of other changes followed before operational trials with a crew provided by the STAT took place at Canjuers from January to March 1986. The AMX40 was judged to be suitable for production, and NBC tests were conducted in the course

BELOW The army's evaluation of the AMX40 proved beneficial not only to GIAT, but also allowed a good amount of experience and data to be collected. The handling of a 120mm weapon, the practical considerations of using combustible case ammunition and the use of advanced fire controls were all factors that needed to be confronted in the ECP. *(Thomas Seignon)*

LEFT Although we cannot make out the full 'W-100 …' registration, its presence as well as the STAT marking indicate that this tank was under evaluation by the French Army at the time this photo was taken. The definitive AMX40 hull shared a certain resemblance to the later Leclerc, and was conceived to offer very high mobility: a power-to-weight ratio around 26hp/tonne, a V12 diesel generating possibly 1,300hp and a weight of under 44 tonnes. *(Jerome Hadacek)*

BELOW The AMX30C2 was the last of several attempts to interest foreign armies in upgrading their AMX30Bs (or AMX30Ss in the case of the Saudis). It was the first to propose cooperation with Kraus-Maffei-Wegman, whose interest in co-developing AMX40 variants with GIAT had been manifested in 1989–90. This cutaway drawing of the hull shows its salient features, including Leopard wheels, not present on the single surviving prototype. *(Nexter Systems/Collection Jerome Hadacek)*

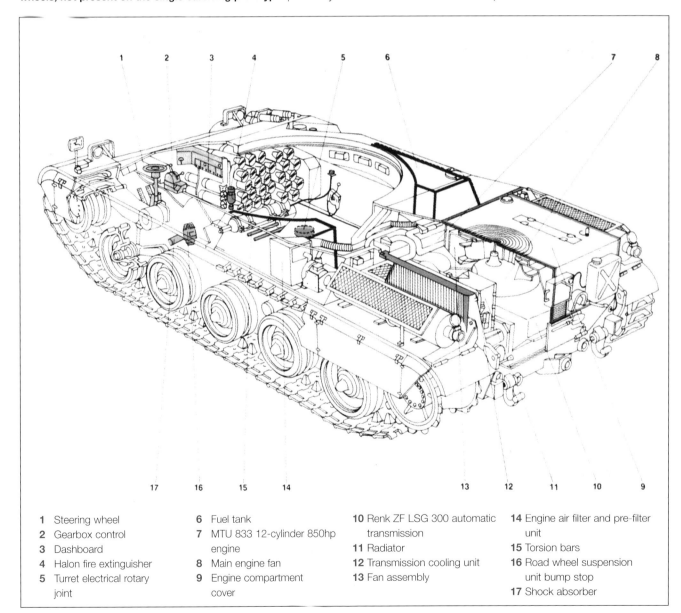

1 Steering wheel	**6** Fuel tank	**10** Renk ZF LSG 300 automatic transmission
2 Gearbox control	**7** MTU 833 12-cylinder 850hp engine	**11** Radiator
3 Dashboard	**8** Main engine fan	**12** Transmission cooling unit
4 Halon fire extinguisher	**9** Engine compartment cover	**13** Fan assembly
5 Turret electrical rotary joint		

14 Engine air filter and pre-filter unit	
15 Torsion bars	
16 Road wheel suspension unit bump stop	
17 Shock absorber	

of 1987. The tracks were replaced with wider 'Chenilles OTAN' double-pin tracks after extensive testing at Biscarosse. Finally, a fourth prototype hull (CH4) was produced with a 1,300hp version of the Poyaud V12X and a new ESM500 transmission. The CH4 hull was able to use the T2 or T3 turret from the second and third prototypes and used both during its trials. The AMX40 P3 was evaluated with a STAT crew in Djibouti in 1987, retaining the AMX30 tracks, which proved inadequate for the mass and power of the tank.

The AMX30 track configuration was retained for the Sharourah trial in Saudi Arabia in 1988, which saw AMX40 P3 and P4 (the CH4 hull and P2 turret) put through their paces through a rocky landscape close to the Yemeni border. Despite this and other concerted efforts to sell the AMX40 to several Middle Eastern countries, the AMX40 matured at the very end of the Cold War. As a consequence, GIAT's priority shifted quickly towards the far more advanced Leclerc programme, and the AMX40 was never purchased by the Saudis or any other customers. In the meantime, the availability of surplus Leopard 2s and the formidable performance of the M1A1 Abrams in 1991 ruined GIAT's hopes of AMX40 export sales. While very much a different tank from the AMX30 and its derivatives, the AMX40 is significant as the last attempt by DEFA's heirs to develop a modern tank in the relatively simple spirit of the old FINABEL 3A5.

An oft-forgotten spin-off of the AMX40 programme was the AMX30C2 upgrade (offered for export by the partnership of GIAT and Krauss-Maffei-Wegmann). This upgrade was offered in 1990 in the hope of interesting foreign armies (especially Saudi Arabia, the United Arab Emirates

and the Qataris) in buying upgraded AMX30Bs, or in upgrading their existing equipment. The options offered included most of the AMX30B2's suspension upgrades, Leopard 1 wheels, a Renk SF LSG 3000 automatic transmission, a German MTU MB833 850hp engine and the GIAT CN-120-G1 120mm gun. There were no takers. As a consequence of mounting the COSTAC system's gun control equipment, the turret bustle was fitted with a very large welded box. The right side of the mantlet was fitted with an armoured sight barbette, presumably equipped with the M571 sight and DIVT16 thermal camera. The Leclerc MBT that eventually emerged in the 1990s to sweep these designs into the past was a weapon of tremendous sophistication, a different approach to tank design better suited to today's professional army.

ABOVE The AMX30C2 prototype survives at Saumur in poor condition. *(Massimo Foti)*

BELOW The scope of trials and test vehicles based on the AMX30 series vehicles is still only partly known, but this image is representative of some of the least commonly acknowledged. Several AMX30B2s were used to test out the best ways to minimise an MBT's signature on different types of battlefield sensory equipment, research which proved valuable to the Leclerc and other programmes. Known collectively as *'Chars Furtives'*, the two seen here were photographed at Mourmelon in 1997. *(Thomas Seignon)*

AMX30 characteristics

	AMX30B	AMX30B2	AMX30B2 Brennus
Produced	1966–85	1982–87 (new) 1987–92 (conversion)	1995–96 (conversion)
Weight combat	36,000kg	37,000kg	38,600kg
Weight unladen	34,000kg	35,000kg	36,600kg
Power to weight	20hp/t	19hp/t	17.8hp/t
Ground pressure	0.77kg/cm2	0.90kg/cm2	0.94kg/cm2
Track width	570mm	570mm	570mm
Track type	single-pin, dead	single-pin, dead or double-pin, live	single-pin, dead or double-pin, live
Maximum width	310cm	310cm	310cm
Hull length	659cm	659cm	659cm
Overall length gun, 12 o'clock	948cm	948cm	948cm
Overall length gun, 6 o'clock	873cm	873cm	873cm
Height of hull	150cm	150cm	150cm
Height to turret roof	229cm	229cm	229cm
Height to cupola searchlight	286cm	286cm	286cm
Trunnion height	181cm	181cm	181cm
Ground clearance	44cm	44cm	44cm
Trackbase	412cm	412cm	412cm
Maximum road speed	60km/h	60km/h	60km/h
Maximum est. cross-country speed	50km/h	50km/h	50km/h
Fuel capacity	970 litres	900 litres	900 litres
Road range (est.)	500km	450km	450km
Fording height (unprepared)	130cm	130cm	130cm
Fording height (without snorkel)	220cm	220cm	220cm
Fording height (with snorkel)	390cm	390cm	390cm
Side slope	17% grade	17% grade	17% grade
Gradient	60% grade	60% grade	60% grade
Vertical obstacle	93cm	93cm	93cm
Trench	290cm	290cm	290cm
Engine	Hispano-Suiza 110	Hispano-Suiza 110	Hispano-Suiza 110 or Mack E9
Engine horsepower (governed to)	710 (680)–2,600rpm	740 (700)–2,600rpm	740 (700) or 750–2,600rpm
Transmission	BV-5-SD (Manual)	ENC-200 (automatic)	ENC-200 (automatic)
Suspension	torsion bar, rear drive, front idler	torsion bar, rear drive, front idler	torsion bar, rear drive, front idler

	AMX30B	AMX30B2	AMX30B2 Brennus
Road wheels	five twin road wheels per side	five twin road wheels per side	five twin road wheels per side
Main armament	CN-105-F1 (DEFA D1512, CN-105-Mle 62)	CN-105-F1 (DEFA D1512, CN-105-Mle 62)	CN-105-F1 (DEFA D1512, CN-105-Mle 62)
Secondary armament	M2 12.7mm HMG or CN-20-F2 (M693) 20mm	CN-20-F2 (M693) 20mm	CN-20-F2 (M693) 20mm
TOP7 cupola machine gun	AN F1 7.62mm	AN F1 7.62mm (armoured)	AN F1 7.62mm (armoured)
105mm stowage, right side hull front	28	28	28
105mm stowage, turret bustle	18	18	18
105mm stowage, turret loader's rack	3	1	1
Coaxial armament ammunition stowage	1,050 12.7mm (turret) or 480 20mm (turret)	480 20mm (turret)	480 20mm (turret)
TOP7 cupola machine gun AN F1 7.62mm	2,050 (including exterior)	2,070 (including exterior)	2,070 (including exterior)
Turret traverse 360 degrees (seconds)	T105, 12 seconds, hydraulic	T105M, 12 seconds, hydraulic	T105M, 12 seconds, hydraulic
Main armament elevation	5.5 degrees/sec, -8 to +20 degrees	5.5 degrees/sec, -8 to +20 degrees	5.5 degrees/sec, -8 to +20 degrees
Secondary armament elevation	mechanical -8 to +40 degrees	mechanical -8 to +40 degrees	mechanical -8 to +40 degrees
Armour type	Cast and rolled homogenous armour	Cast and rolled homogenous armour	Cast and rolled homogenous armour + ERA
Turret front	Cast, 80mm	Cast, 80mm	Cast, 80mm + ERA
Turret sides	Cast, 41mm	Cast, 41mm	Cast, 41mm + ERA
Turret roof	Cast, 20mm	Cast, 20mm	Cast, 20mm + ERA
Turret rear	Cast, 50mm	Cast, 50mm	Cast, 50mm
Gun mantlet	Cast, 81mm	Cast, 81mm	Cast, 81mm + ERA
Glacis	Cast, 79mm	Cast, 79mm	Cast, 79mm + ERA
Hull sides	RHA 57mm	RHA 57mm	RHA 57mm
Hull floor	RHA 15mm	RHA 15mm	RHA 15mm
Hull sides engine deck	RHA 30mm	RHA 30mm	RHA 30mm
Hull roof	RHA 15mm	RHA 15mm	RHA 15mm
Hull rear	RHA 30mm	RHA 30mm	RHA 30mm
Fire controls	Coincidence rangefinder, gunner's telescopic and periscopic sights, infrared night sights	COTAC system with laser rangefinder, DIVT13 low-light television camera, or DIVT16 thermal television camera	COTAC system with laser rangefinder, DIVT16 or DIVT18 thermal television camera

AMX30 units

French Army AMX30 Regiments (Arme Blindée Cavalrie)	Brigade (1967)	Division Mécanisée (Type 1967)	Division Blindée (Type 1977)
1e Régiment de Cuirassiers	1e Brigade Mécanisée	1e Division Mécanisée	1e Division Blindée
501e Régiment de Chars de Combat	2e Brigade Mécanisée	8e Division Mécanisée	2e Division Blindée
5e Régiment de Cuirassiers	3e Brigade Mécanisée	1e Division Mécanisée	5e Division Blindée
2e Régiment de Cuirassiers	5e Brigade Mécanisée	3e Division Mécanisée	5e Division Blindée
2e Régiment de Dragons	6e Brigade Mécanisée	7e Division Mécanisée	6e Division Blindée
30e Régiment de Dragons	7e Brigade Mécanisée	7e Division Mécanisée	dissolved in Sept 1978
3e Régiment de Cuirassiers	8e Brigade Mécanisée	7e Division Mécanisée	4e Division Blindée
503e Régiment de Chars de Combat	10e Brigade Mécanisée	4e Division Mécanisée	10e Division Blindée
6e Régiment de Dragons	11e Brigade Mécanisée	1e Division Mécanisée	1e Division Blindée
12e Régiment de Cuirassiers	12e Brigade Mécanisée	3e Division Mécanisée	3e Division Blindée
6e Régiment de Cuirassiers	14e Brigade Mécanisée	8e Division Mécanisée	2e Division Blindée
2e Régiment de Chasseurs	15e Brigade Mécanisée	4e Division Mécanisée	4e Division Blindée
4e Régiment de Cuirassiers	16e Brigade Mécanisée	4e Division Mécanisée	6e Division Blindée
507e Regiment de Chars de Combat	Saumur Training Regiment	Saumur Training Regiment	Saumur Training Regiment
1e Régiment de Dragons	NA	NA	7e Division Blindée
5e Régiment de Dragons	NA	formed from 30e Dragons 1978–79	7e Division Blindée
4e Régiment de Dragons	NA	NA	10e Division Blindée
11e Régiment de Cuirassiers	NA	NA	NA
1e Régiment de Chasseurs	NA	NA	NA
3e Régiment de Chasseurs	NA	NA	NA
11e Régiment de Chasseurs	NA	NA	NA
3e Régiment de Dragons	NA	NA	3e Division Blindée
		1967 System: five Divisions Mécaniséés in two Corps d'Armée	1977 System: eight Divisions Blindées in three Corps d'Armée

Year equipped with AMX30B	Division Blindées 1984	After 1990
1969 RC54	1e Division Blindée (RC70)	Amalgamated with 11e Régiment de Cuirassiers 1999 (RC80)
1967–68 RC54	2e Division Blindée	Amalgamated with 503e Régiment de Chars de Combat 1994 (RC80)
1969 RC54	5e Division Blindée	Dissolved in 1992
1971 RC54	5e Division Blindée	Dissolved in 1991
prior to 1980 RC54	2e Division Blindée	Last regiment equipped with AMX30B
1968 RC54		
1973 RC54	7e Division Blindée	Dissolved in 1998
1966–67 RC54	10e Division Blindée	Amalgamated with 501e Régiment de Chars de Combat 1994
1972 RC54	1e Division Blindée (RC70)	Dissolved in 1992
1970 RC54	3e Division Blindée	Amalgamated with 6e Régiment de Cuirassiers 1994 (RC80)
1970 RC54	2e Division Blindée	Amalgamated with 12e Régiment de Cuirassiers 1994 (RC80)
1970 RC54	10e Division Blindée	Amalgamated with 1e Régiment de Chasseurs 1998 (RC80 Brennus)
1970 RC54	5e Division Blindée (RC70)	Dissolved in 1997
1968 partial equipment on training establishment	12e Division Légère Blindée	Dissolved in 1997
1981 RC54	7e Division Blindée	Dissolved in 1997
1979 RC54	7e Division Blindée	Dissolved in 2003
1981 RC54	10e Division Blindée	Dissolved in 1994 after service in Operation Daguet
1981 RC54	14e Division Légère Blindée	Amalgamated with 1e Régiment de Cuirassiers 1999 (RC80)
1981 RC54	14e Division Légère Blindée	Amalgamated with 2e Régiment de Chasseurs 1998 (RC80 Brennus)
recreated 1981 RC54	12e Division Légère Blindée	Dissolved in 1997
1984 RC54	Forces Françaises à Berlin	Dissolved in 1994
1976 RC54	3e Division Blindée	Dissolved in 1997
Approximate number of tanks in service by 1986: 1,200 in ABC (including 200 AMX30B2s)	1984 System: six Divisions Blindées and two Divisions Légères Blindées in three Corps d'Armée plus one DLB in the Force d'Action Rapide	RC54: 54 tanks in 4 squadrons of 13 tanks plus regimental command tanks, RC52: 52 tanks in 3 squadrons of 17 tanks plus regimental command tanks, RC70: was a reinforced RC52 with an additional squadron and command tank. RC80 was regimental structure developed for the Leclerc with 80 tanks in 2 regimental groups of 40 tanks (3 squadrons of 13 tanks plus 1 command tank)

Régiments d'Infanterie Mécanisé (RIMECA)	Division Blindée (1984)	FFA or FRANCE Metropole	Year equipped with AMX30B	Number of tanks
8e Groupe de Chasseurs	1e Division Blindée	FFA	1982–86	Two compagnies or one escadron (max 16)
16e Groupe de Chasseurs	1e Division Blindée	FFA	1982–86	Two compagnies or one escadron (max 16)
5e Régiment d'Infanterie	2e Division Blindée	FRANCE	1985–87	One compagnie (10 tanks)
Régiment de Marche du Tchad	2e Division Blindée	FRANCE	1985–87	One compagnie (10 tanks)
19e Groupe de Chasseurs	3e Division Blindée	FFA	1982–86	Two compagnies or one escadron (max 16)
42e Régiment d'Infanterie	3e Division Blindée	FFA	1982–86	Two compagnies or one escadron (max 16)
2e Groupe de Chasseurs	4e Division Blindée	FFA	1982–86	Two compagnies or one escadron (max 16)
24e Groupe de Chasseurs	4e Division Blindée	FFA	1982–86	Two compagnies or one escadron (max 16)
35e Régiment d'Infanterie	7e Division Blindée	FRANCE	1985–87	One compagnie (10 tanks)
170e Régiment d'Infanterie	7e Division Blindée	FRANCE	1985–87	One compagnie (10 tanks)
1e Groupe de Chasseurs	10e Division Blindée	FRANCE	1985–87	One compagnie (10 tanks)
150e Régiment d'Infanterie	10e Division Blindée	FRANCE	1985–87	One compagnie (10 tanks)
3e Régiment d'Infanterie*	14e Division Légère Blindée	FRANCE	1985–89	Maximum of 12 tanks
* Training regiment, incomplete equipment				Number of AMX30B tanks in RIMECA by 1991: 182–200 vehicles

(Charles Beaudouin)

Production chart of AMX30B2s

Serial	Chassis No.	Radio fit	Turret No.	Note	Delivery	Unit (where applicable)
AMX30B2 Batch 1 GIAT Order 6354.00 1980–82 (8 Tanks)						
6814-0099	1	Platoon Command	5001	First AMX30B2 Prototype	25.11.1981	AMX/APX evaluation
6814-0116	2	Platoon Command	5002		17.12.1981	STAT Satory
6824-0100	3	Platoon Command	5003		25.05.1982	STAT Satory then 503e RCC
6824-0159	4	rank	5004		23.08.1982	503e RCC
6824-0174	5	rank	5005		11.10.1982	ETBS Bourges
6824-0206	6	rank	5006		16.11.1982	503e RCC
6824-0209	7	rank	5007		22.11.1982	503e RCC
6834-0021	8	rank	5008		20.12.1982	503e RCC
AMX30B2 Batch 2 GIAT Order 6457.00 1983–84 (54 tanks)						
6834-0034	9	rank	5009		18.03.1983	CPCIT Canjuers
6834-0085	11	rank	5011		21.04.1983	503e RCC
6834-0090	10	Platoon Command	5013		14.06.1983	503e RCC
6834-0035	15	rank	5010		14.06.1983	EAABC Saumur
6834-0094	12	rank	5012		30.06.1983	503e RCC
6834-0096	13	rank	5016		30.06.1983	503e RCC
6834-0097	14	Regimental Command	5014		30.06.1983	503e RCC
6834-0099	18	rank	5019		14.06.1983	503e RCC
6834-0107	16	Platoon Command	5015		08.07.1983	EAABC Saumur
6834-0114	19	Platoon Command	5027		24.08.1983	CPCIT Canjuers
6834-0115	22	rank	5028		31.08.1983	503e RCC
6834-0129	17	rank	5023		31.08.1983	503e RCC
6834-0141	21	rank	5024		14.10.1983	503e RCC
6834-0142	25	Regimental Command	5022		14.10.1983	503e RCC
6834-0140	20	rank	5017		14.10.1983	503e RCC
6834-0144	23	Platoon Command	5018		14.10.1983	503e RCC
6834-0153	31	rank	5020		14.10.1983	503e RCC
6834-0151	26	rank	5029		14.10.1983	503e RCC
6834-0152	27	rank	5030		14.10.1983	503e RCC
6834-0161	29	rank	5026		23.11.1983	503e RCC
6834-0162	28	Platoon Command	5032		23.11.1983	503e RCC
6834-0163	33	Platoon Command	5024		07.12.1983	503e RCC
6834-0172	29	rank	5039		23.11.1983	503e RCC
6834-0173	35	Platoon Command	5025		23.11.1983	503e RCC
6834-0179	30	Platoon Command	5036		06.01.1984	503e RCC
6834-0180	32	rank	5033		07.12.1983	503e RCC
6834-0187	43	Platoon Command	5037		07.12.1983	503e RCC
6834-0188	38	Regimental Command	5031		07.12.1983	503e RCC
6834-0191	34	Platoon Command	5044		23.12.1983	503e RCC
6834-0192	36	rank	5040		23.12.1983	503e RCC
6834-0200	42	rank	5045		23.12.1983	503e RCC
6834-0201	44	rank	5041		23.12.1983	503e RCC
6834-0186	41	Regimental Command	5038		06.01.1984	503e RCC
6844-0001	40	rank	5035		24.02.1984	4e RD
6844-0002	37	rank	5042		24.02.1984	4e RD
6844-0003	48	rank	5047		24.02.1984	4e RD

Serial	Chassis No.	Radio fit	Turret No.	Note	Delivery	Unit (where applicable)
6844-0013	39	Regimental Command	5050		24.02.1984	4e RD
6844-0014	45	Platoon Command	5049		21.02.1984	4e RD
6844-0015	46	rank	5053		21.02.1984	4e RD
6844-0016	47	rank	5039		21.03.1084	4e RD
6844-0017	52	Platoon Command	5048		21.03.1984	4e RD
6844-0023	50	rank	5054		06.04.1984	4e RD
6844-0024	54	rank	5055		06.04.1984	4e RD
6844-0028	51	rank	5052		21.03.1984	4e RD
6844-0037	49	rank	5056		15.05.1984	4e RD
6844-0038	53	Platoon Command	5057		09.05.84	CPCIT Canjuers
6844-0039	55	rank	5046		16.05.1984	4e RD
6844-0040	56	rank	5059		16.05.1984	4e RD
6844-0048	58	rank	5061		09.05.1984	4e RD
6844-0049	60	Platoon Command	5058		16.05.1984	4e RD
6844-0052	59	Platoon Command	5060		15.05.1984	4e RD
6844-0056	62	Regimental Command	5043		04.06.1984	ETAS Angers
6844-0115	67	Regimental Command	5069		13.07.1984	4e RD
6844-0122	63	Platoon Command	5065		12.09.1984	4e RD

AMX30B2 Batch 3 GIAT Order 6636,00 1984–85 60 Tanks						
6844-0053	61	rank	5062		09.05.1984	CPCIT Canjuers
6844-0062	57	rank	5063		06.06.1984	EAABC Saumur
6844-0063	65	rank	5071		03.07.1984	EAABC Saumur
6844-0123	64	rank	5051		25.09.1984	CPCIT Canjuers
6844-0125	68	Platoon Command	5066		12.09.1984	4e RD
6844-0132	69	Platoon Command	5073		16.10.1984	4e RD
6844-0133	70	rank	5070		16.10.1984	4e RD
6844-0126	71	rank	5067		16.10.1984	4e RD
6844-0139	66	rank	5076		22.11.1984	507e RCC
6844-0140	74	Platoon Command	5074		28.11.1984	2e RCh
6844-0150	72	Regimental Command	5079		22.11.1984	2e RCh
6844-0151	80	rank	5072		22.11.84	507e RCC
6844-0152	79	rank	5073		28.11.1984	2e RCh
6844-0156	77	Platoon Command	5075		22.11.1984	ESAM Bourges
6844-0155	82	rank	5082		28.11.1984	2e RCh
6844-0162	73	Platoon Command	5080		09.01.1985	2e RCh
6844-0157	81	rank	5064		09.01.1985	2e RCh
6844-0164	83	rank	5077		09.01.1985	2e RCh
6844-0168	75	Platoon Command	5081		09.01.1985	2e RCh
6844-0169	76	Platoon Command	5085		30.01.1985	2e RCh
6844-0170	85	rank	5078		09.01.1985	2e RCh
6844-0176	78	rank	5068		09.01.1985	2e RCh
6844-0177	84	rank	5086		09.01.1985	2e RCh
6844-0178	86	rank	5084		09.01.1985	2e RCh
6854-0001	87	Platoon Command	5091		30.01.1985	2e RCh
6854-0002	88	rank	5087		30.01.1985	2e RCh
6854-0012	90	rank	5090		13.03.1985	2e RCh
6854-0016	91	rank	5089		13.03.1985	2e RCh
6854-0013	92	Platoon Command	5092		13.03.1985	2e RCh
6854-0023	93	rank	5093		30.04.1985	2e RCh
6854-0024	95	rank	5094		30.04.1985	2e RCh
6854-0025	96	rank	5097		15.03.1985	CIABC CARPIAGNE
6854-0011	89	rank	5095		13.03.1985	2e RCh
6854-0039	94	Platoon Command	5100		29.05.1985	2e RCh
6854-0040	98	rank	5102		15.05.1985	CIABC CARPIAGNE
6854-0062	100	Platoon Command	5107		20.06.1985	2e RCh
6854-0048	104	rank	5098		15.05.1985	CIABC CARPIAGNE
6854-0051	99	rank	5109		03.10.1985	2e RC
6854-0063	103	rank	5101		27.06.1985	2e RC
6854-0068	102	rank	5105		05.09.1985	2e RC
6854-0067	97	rank	5112		05.06.85	2e RC
6854-0076	101	rank	5108		05.06.85	2e RC
6854-0077	105	Regimental Command	5099		05.06.85	2e RC

Serial	Chassis No.	Radio fit	Turret No.	Note	Delivery	Unit (where applicable)
6854-0080	107	rank	5106		10.06.85	2e RC
6854-0081	108	rank	5113		10.06.85	2e RC
6854-0082	112	Platoon Command	5103		10.06.85	2e RC
6854-0090	106	Platoon Command	5096		01.07.85	2e RC
6854-0091	109	rank	5114		01.07.85	2e RC
6854-0092	110	Regimental Command	5110		01.07.85	2e RC
6854-0100	114	Regimental Command	5115		15.07.85	2e RC
6854-0101	115	Platoon Command	5117		15.07.85	2e RC
6854-0102	116	Platoon Command	5116		15.07.85	2e RC
6854-0106	120	rank	5118		26.08.85	2e RC
6854-0110	111	Platoon Command	5088		11.09.85	11e RC CIABC CARPIAGNE
6854-0116	113	Platoon Command	5122		16.09.85	2e RC
6854-0117	119	Regimental Command	5120		16.09.85	2e RC
6854-0130	117	Regimental Command	5128		23.10.85	2e RC
6854-0131	122	Regimental Command	5121		23.10.85	2e RC
6854-0124	118	rank	5113		29.10.85	2e RC
6854-0139	121	rank	5123		29.10.85	2e RC

AMX30B2 BATCH 4 GIAT ORDER 64740.00 1986–87 (44 tanks)						
6854-0143	125	rank	5126		18.11.85	5e RC
6854-0139	123	rank	5123		25.11.85	5e RC
6854-0146	124	rank	5124		25.11.85	5e RC
6854-0147	127	rank	5130		25.11.85	5e RC
6854-0157	126	rank	5105		11.12.85	5e RC
6854-0158	132	rank	5127		11.12.85	5e RC
6864-0073	130	Regimental Command	5126		18.11.85	STAT SATORY
6864-0074	133	Platoon Command	5129		15.01.86	5e RC
6864-0079	129	rank	5124		10.02.86	5e RC
6864-0080	131	rank	5130		10.02.86	5e RC
6864-0077	128	Platoon Command	5105		09.04.86	5e RC
6864-0089	135	rank	5127		09.04.86	5e RC
6864-0090	138	rank	5132		09.04.86	5e RC
6864-0094	134	rank	5133		29.04.86	5e RC
6864-0081	136	Platoon Command	5134		23.04.86	5e RC
6864-0095	133	rank	5125		23.04.86	5e RC
6864-0098	137	rank	5111		23.09.86	4e RC
6864-0099	140	Platoon Command	5137		24.11.86	4e RC
6864-0104	142	rank	5145		26.05.86	4e RC
6864-0112	141	rank	5139		18.11.86	4e RC
6864-0105	143	rank	5143		03.06.86	4e RC
6864-0108	144	rank	5147		03.06.86	4e RC
6864-0111	146	rank	5146		09.06.86	4e RC
6864-0103	149	Platoon Command	5148		30.06.86	4e RC
6864-0113	159	rank	5162		20.10.86	4e RC
6864-0117	151	Platoon Command	5153		07.07.86	4e RC
6864-0115	148	Platoon Command	5151		17.07.86	4e RC
6864-0128	150	rank	5131		17.07.86	4e RC
6864-0126	154	Regimental Command	5149		17.07.86	4e RC
6874-0046	155	rank	5156		18.08.86	4e RC
6874-0047	152	rank	5152		11.09.86	4e RC
6864-0146	147	Regimental Command	5164		08.09.86	4e RC
6874-0048	160	Platoon Command	5160		16.09.86	4e RC
6874-0049	157	Platoon Command	5140		13.10.86	4e RC
6864-0147	153	rank	5154		20.10.86	4e RC
6874-0045	145	rank	5144		27.01.87	4e RC
6874-0050	156	Platoon Command	5155		19.01.87	4e RC
6874-0051	143	Platoon Command	5159		21.01.87	4e RC
6874-0056	158	rank	5158		27.01.87	3e RD
6874-0052	161	Regimental Command	5157		27.01.87	4e RC
6874-0057	162	rank	5150		05.02.87	3e RD
6874-0064	165	rank	5163		18.02.87	3e RD
6874-0063	164	Platoon Command	5165		02.03.87	3e RD
6874-0067	166	rank	5164		02.03.87	3e RD

French Army AMX30 inventory

Type (New Build)	Year adopted	Total produced French Army	*= estimated, **= includes prototypes
AMX30B	1966	1173	*
AMX30D	1970	148	
AMX30P	1973	40	
AU F1 (H and T)	1978	227	*
AU F1 TM	1991	24	
AMX30R (Roland 1)	1978	83	
AMX30R (Roland 2)	1981	98	
AMX30B2	1981	166	**
EBG	1987	71	

AMX30 Conversions	Year	Total converted French Army	Original vehicle
AMX30B2 (GIAT)	1987	236	AMX30B
AMX30 B2 (Gien)	1988	257	AMX30B*
AMX30 EBD	1990	5	AMX30B
AMX30B2 Brennus	1995	81	AMX30B2
AMX30B2 DT	1994	5	AMX30B2 (turret AMX30B)
AMX30B2 DT	1997	10	AMX30B2
AU F1 (TA)	2005	106	AMX30B2
EBG Valorisé	2006	30	EBG
EBG Valorisé SPDMAC	2006	12	EBG

Estimated MBT inventory	1991	514	AMX30B
		659	AMX30B2

Appendix 5

AMX30 exports

Country	Type	Year	Total	Notes
Bosnia	AMX30B (ex-United Arab Emirates)	1998–2000	50	date, number unconfirmed
Chile	AMX30B	1980	21	SIPRI
	AMX30B (ex-France)	1998	21	SIPRI
	AMX30B (ex-France)	2000	10	unconfirmed
Cyprus	AMX30B2	1987	15	SIPRI
	AMX30B2	1989–91	35	SIPRI
	AMX30D	1989	1	SIPRI
	AMX30B (ex-Greece)	1992	40 *	40 and 52 reported
	AMX30D (ex-Greece)	1992	4	unconfirmed
Greece	AMX30B	1970–71	55 (60)	SIPRI, 60 reported elsewhere
	AMX30B	1974–78	130	SIPRI
	AMX30D	1976–78	14	SIPRI
Iraq	AMX30 155 AU GCT	1981–85	85	SIPRI
	AMX30D	1981	5	SIPRI
	AMX30R Roland 2	1981–85	14	SIPRI
Kuwait	AMX30 155 AU GCT	1989	18	SIPRI
Nigeria	AMX30R Roland 2	1989*	18	unconfirmed
Qatar	AMX30B	1975	24	SIPRI
	AMX30D	1977	1	SIPRI
	AMX30B (ex-France)	1990–2000	10 (20)	SIPRI, 20 reported elsewhere
Saudi Arabia	AMX30S	1976–85	290	SIPRI
	AMX30D	1976–81	57	SIPRI
	AMX30H Poseur de Pont	1979–81	12	SIPRI
	AMX30 155 AU GCT	1980–85	51 (63)	63 from SIPRI
	AMX30R Roland 2	1981–85	52	unconfirmed, possibly less
	AMX30 SA Shahine (Surveillance)	1980–89	12	unconfirmed, possibly more
	AMX30 SA Shahine (Missile)	1980–89	24	unconfirmed, possibly more
	AMX30 30mm Bitube DCA	1978–85	51	SIPRI
Spain	AMX30B	1969	19	not counted in SIPRI total
	AMX30E	1974–79	180	license built, SIPRI
	AMX30D	1974–78	10	unconfirmed possibly more
	AMX30E	1980–83	100	license built, SIPRI
	AMX30R Roland 1	1980	9	SIPRI
	AMX30R Roland 2	1981–83	9	SIPRI
United Arab Emirates	AMX30B	1974	64	SIPRI
	AMX30D	1974	4	SIPRI
Venezuela	AMX30B	1973–74	81	SIPRI
	AMX30D	1973–74	4	SIPRI

Total (estimated) for all AMX30 variants produced for export:
1,194 vehicles, with an additional 280 AMX30E built under license in Spain.

Glossary of terms and abbreviations

ABS: *Arsenal de Bourges*
AMX: *Ateliers d'Issy les Moulineaux*
APX: *Ateliers de Puteaux*
ARE: *Ateliers de Roanne*
ARL: *Ateliers de Rueil*
ATS: *Ateliers de Tarbes*
CAFL: *Compagnie des Ateliers et Forges de la Loire*
CN: *Canon* (which translates to cannon)
DEFA: *Direction des Etudes et Fabrication d'Armement*
DGA: *Direction Générale de l'Armement*
DTAT: *Direction Technique de l'Armement Terrestre*
EFAB: *Etablissement de Fabrication d'Armement de Bourges*
EMAT: *Etat-Major de l'Armée de Terre*
GAP: *Groupe auxiliaire de puissance* (which translates to auxiliary powertrain or generator)
GIAT: *Groupement des Industriels de l'Armement Terrestre*
IGA: *Ingénieur Général de l'Armement*
MAS: *Manufacture d'Armes de Saint-Étienne*
OB: *Obus* (which translates to projectile)
SAVIEM: *Société Anonyme de Véhicules Industriels et Mécanique*
SOPELEM: *Société Optique de Précision ELEctronique et Mécanique*
STA: *Section Technique de l'Armement*
STAT: *Section Technique de l'Armement de Terre*

Further reading.

Brindeau, P. and Maillet, I.G.A., *Tome 7: Materiels du Genie* (Comité pour l'Histoire de l'Armement Terrestre (COMHART), Centre des Hautes Études de l'Armement, Division Histoire, DGA, Paris, 2007)

Jeudy, Jean-Gabriel, *Chars de France* (Editions Techniques pour l'Automobile et l'Industrie, Paris, 1997)

Robineau, B., *Tome 5: Relations Internationales* (Comité pour l'Histoire de l'Armement Terrestre (COMHART), Centre des Hautes Études de l'Armement, Division Histoire, DGA, Paris, 2003)

Robinson, M.P., *AMX30: Char de Bataille, Volume 1* (Kagero, Lublin, 2014)

Robinson, M.P., *AMX30: Char de Bataille, Volume 2* (Kagero, Lublin, 2015)

Robinson, M.P., *The AMX30 Family* (Kagero, Lublin, 2015)

Robinson, M.P. and Cany, J., *AMX30: l'Épopée d'un Char Francais* (Editions Cany, St. Pierre, 2015)

Tauzin, M. and Marest, M., *Tome 9: L'Armement de Gros Calibre* (Comité pour l'Histoire de l'Armement Terrestre (COMHART), Centre des Hautes Études de l'Armement, Division Histoire, DGA, Paris, 2008)

Touzin, P., *Les Véhicules Blindés Français 1945–1977* (Editions EPA, Paris, 1978)

Index

Ammunition 44, 46, 48, 55, 102, 124, 131
  APDS 24
  APFSDS 47, 57, 85
  HEAT 46, 48, 84, 89-90
    Obus-G (OCC-F1) 105mm 24, 40, 56, 83-84, 89-90, 93, 150
  hollow-charge 14-15, 40, 83, 89, 122
  kinetic energy (KE) 41, 46, 54-56, 83-84, 90, 95, 153
  TIAB 121
Ammunition stowage 47-48, 50, 55, 68, 87, 90, 102, 136
AMX (*Ateliers d'Issy les Moulineaux*) bureau 10, 27, 68
  APX (*Atelier de Puteaux*) design team 30-31, 171-173

AMX30
  definition vehicles 27-29, 68
  prototypes 11, 15-20, 25-26, 68upgrades and rebuilds 7-8, 34, 38, 46-47, 51, 54, 59, 125, 154, 158
variants 7-8, 27, 38, 109-147
AMX30-based engineers vehicles 139-147
  ENFRAC river-crossing vehicle prototype 147
  Pinguely crane-equipped prototype 147
AMX30A 22, 24
  preseries 11, 18, 20
  prototypes 20-22, 27, 29
AMX30 ACRA missile launcher 27, 30, 33, 46, 171-172
AMX30B 5, 20, *et seq.*
AMX30B2 5-6 *et seq.*
  conversions 7, 53, 55-57, 59-61, 129, 144-145, 161, 164, 168
  definition prototype 53
AMX30B2 Brennus 7, 55-56, 59-61, 79, 146, 155, 168, 170
  demonstrator 59, 61
AMX30B2 DT mine-clearing vehicles 144-147
  mine roller systems 144-146
  Ramta mine ploughs 144-145, 147
AMX30 Bitube 30mm DCA 136-138, 159
AMX30C2 upgrade 175
AMX30D recovery vehicle 8, 30, 33, 38, 45, 59, 79, 110-115, 139, 154, 157-160
  crane 105, 110, 112
  lifting jig 112
  recovery and repair equipment 113
  winch 111-112
AMX30E 151-153; AMX30EM2 153-158 ; AMX30ER1 151, 153
AMX30EBD 144
  mine rollers 144
AMX30 EBG 59, 71, 81, 109, 139-143
  boring drill 140
  demolition charge launcher 140-141
  hydraulic arm 139-143
  multiple rocket launcher 142
AMX30 EBG SDPMAC 142-143
  mine-clearing device 142-143
AMX30H bridge-layer 33, 121, 143-144, 158
  Class 50 scissor bridge 144
AMX30P Pluton erector-launchers 33-34, 38, 71, 114-119, 139
  crane 115, 117
AMX30R Roland 1 and 2 SAM 33, 38, 59, 72, 119, 132-135, 137, 159
AMX30S 50, 143, 149, 159-161, 169-170, 174
AMX30SA radar tracking vehicle 135, 138
AMX30 Shahine launcher 134-136
AMX30V 158-159
AU F1 (AU F1 H) self-propelled gun (AMX30 155 GCT) 28, 34, 50, 59, 70, 79, 119-, 121-132, 137-138, 159-160, 169-170
  automatic loading system 121, 123-124, 126, 128, 130-131
  gunnery aids 130-131
  inertial navigation 131
  mobile fire command centres 131
AU F1 T 72, 126-129; AU F1 TA 8, 81, 128-130; AU F1 TM 127
AU F2 (cancelled) 128, 130
AMX32 33, 41-51, 56-57, 70, 84, 87, 152, 172-173

Anti-aircraft gun (SPAAG) 137-138
Armament, main 7, 28, 33, 83-97
  CN-105-F1 105mm (D1512) 24, 30, 39, 48, 53, 55, 83-86, 90, 93, 95, 101, 103, 160, 166
  recoil system 101-102

CN-105-G1 89
CN-105-57 105mm 14, 84-85
CN-120-G1 49-50
CN-155-F1 (CN-155-GCT) 119, 121-124, 128, 130
D1507 105mm 84
HS831 automatic cannon 136
L7 105mm 84, 150; L7A3 151
PAK 43/41 105mm 84
Rheinmetall L44 120mm 48-49
smoke grenade launchers 57, 63, 104, 155, 166, 173
Armament, secondary 49 86-89, 101
AN F1 7.62mm machine gun 90, 97, 113, 126, 140-142
Browning M2 12.7mm 86, 131, 154
CN-20-F1 86
CN-20-F2 (M621) 51, 77, 86-89, 167, 169-170
blast shield 170
DTAT M693 20mm cannon 86
HS820 20mm 87
12.7mm machine gun 42, 77, 86-89, 104, 131, 155, 157, 160
Armoured personnel carrier (APC) AMX13 VTT 21, 150
Armour protection 16, 26, 30, 38, 41, 46, 53, 63, 116, 156
angles 61
appliqué 143
explosive reactive (ERA) 7, 59-61, 63, 144, 146, 155-156, 158
frontal 20
GIAT BR G2 155
hardness 46
laminate 40
mounting stubs 60
rolled homogeneous (RHA) 61
SABBLIR 155-158
spaced 46-47
thickness 19-20, 61, 92, 124
welded 124
Auxiliary power units (GAPs) 115, 118, 125-127, 130, 134-137

Bastille Day parades 129; 1967 35-36; 1968 35; 1980 41, 1991 113, 168; 2014 109
Battle of Khafji 161, 169
Beaudouin, Gén de Div Charles 6-7
Beaudouin, Jean 6
Belgian Army 23
British Army 127
Bundeswehr 14, 144

Carrougeau, Ing Gén Maurice 55
Centauro armoured car 156
Chassis 15-16, 18, 27-28, 45, 56, 71, 121, 125, 130, 132, 136-139, 141-142
Cold War, end of 8, 30, 40, 44, 59, 81, 110, 128, 130, 134, 158, 161, 164, 175
Colli, Col Delli 23
Colour schemes and camouflage 11, 41, 55-56, 58, 113-114, 124, 150, 151, 157, 166, 168
markings 26, 43, 174
mud 41, 154
Communications 76-77, 119, 127-129
antennae 26, 28, 102, 127, 129, 151
ATILA battery command 130-131
radios and wireless sets 24, 54, 67, 76-79, 97, 100, 102, 129, 131, 151
Compressed air system 102, 121
Controls 7, 80, 91, 95, 103, 106-107
commander overriding gunner 95
Crew 5, 8, 43, 104
AMX30D 111
AMX30P 114-115, 118, 124
AU F1 130-132
collective spirit 7
commanders 7, 23, 45, 78, 80, 90-92, 95, 97, 100-101, 104, 130
conscripts 44, 164
drivers 36-38, 93, 104-107, 130
inexperienced 37, 45, 69
escape 44, 105
gun commander 130-131
gun layer 130
gunners 17, 23, 37, 49-50, 78, 88, 92-93, 95, 97, 101, 104, 171
loaders-wireless operators 76, 78, 93, 102-103, 129-131
missile operator 134
mounting the tank 78

personal kit 124
second 134
Spanish 154
training 28-29, 35-37, 44, 69, 77-79, 95, 134, 151, 153, 168
Crew positions 5, 23
commander's 96
driver's 16, 21, 23, 29, 54, 60, 66, 107
gunner's 42, 96, 128
loader's 17, 107, 126, 154
Cupola
ammunition tray 44
commander's 17, 21, 49, 95, 124
SAMM S401 28-29; S470 21, 67
TOP7 7, 27-29, 43, 90, 93, 97, 111, 113, 151
Cypriot National Guard 157-158

Decontamination procedure 80-81
DEFA 9-10, 14-15, 23, 27, 55, 89-90, 92, 110, 136, 150, 175
De Gaulle, Charles 14-15
DGA 28, 48
Doin, Gén 27
DTAT 27-28, 30, 34, 55, 68, 110, 151

Electrics and electronics 28, 55, 57, 99, 103, 115-116, 118-119, 125-126, 134-135, 137
battery compartments 28, 59-60, 77, 79, 107, 125
Engine compartment 16, 60, 105-106, 152
fire extinguisher 69
Engine cooling system 68
radiators 105, 107
Engines 45, 62, 122
AVDS 190 152; 1790-2A and -2C diesel 152, 158
Daimler-Benz 18
Hispano-Suiza V12 HS-110 multi-fuel 23, 48, 51, 54, 58, 61, 68-69, 106, 135, 152, 158
Mack E9 V8 diesel 7, 59-61, 129, 142, 170
Maybach 10-11
MTU MB833 diesel 152, 154, 173, 175
Poyaud V12X diesel 172-173, 175
SAVIEM licensed HS-110 68
SOFAM petrol 11, 15-16, 18, 20, 23, 68
ENOSA 152, 155
Entry into service 7-8, 35-41, 60-61, 63, 104, 133, 161
first public appearances 35, 53
re-equipping regiments and squadrons 36-37, 54
Erprobungstelle 91 23
ETAT 171
Euromissile 132
Exhaust system 15, 17, 25, 33, 71, 105, 125, 130, 147
Export market 7, 30, 34, 47, 50, 125-126, 134, 138, 149-161, 172, 174
Chile 159
Cyprus 157-158
Greece 157
Iraq 50, 126-127
Kuwait 128
Qatar 160-161, 175
Saudi Arabia 50, 125-127, 134, 136, 138, 144, 159-160, 175
Spain 151-157
United Arab Emirates 50, 160, 175
unsuccessful deals 150
Venezuela 158-159

Fiandini, Col 23
Fire control systems 5, 38, 41, 43, 46, 51, 54, 85, 93, 97, 103, 124, 126-127, 134, 153
ATLAS 128-131
computerised 46, 125
COTAC 46, 95-97, 172, 175
gun stabilisation system 158
missile 116, 118, 137
radar-guided 131
SABCA 158
simulator 41
Firing 86, 88, 92
rate of fire 95, 119, 121, 128
Franco, Gen 150
Franco, Gen 151
French Air Force 27
French Army – throughout
AMX30 units 180-182
Arme Blindée Cavalerie (ABC) 6, 8, 10-11, 35-37, 40, 55, 164, 170-171
Gien workshops 8, 34, 57, 59, 110

Mechanised Infantry 42
reduction 58-59, 134
rejects AMX32 47
Section Technique d'Artillerie, St Cloud 84
French nuclear capability 15, 114-115
Colomb-Béchar Accord 15
Fuel system and tanks 69, 106, 158
consumption 26, 69
barrels 57

GIAT Industries 27, 33, 40-41, 43, 46, 50-51, 57, 59, 61, 71, 110, 119, 122, 125-126, 130, 132-133, 135, 138, 143-144, 146, 150, 158-159, 161, 164, 173-174
Gulf War (First, 1990-91) 56-57, 60, 113, 161, 165-170

Hatches 16-17, 20, 60, 77-79, 88, 93, 103, 115, 124, 135, 151, 155
Hellenic Army 150, 157-158
Huberdeau, Lt-Col 23, 35
Hulls 16, 20-21, 28, 30-31, 45-46, 51, 56, 66, 72, 85, 111, 116, 125-127, 132, 135
glacis 16, 20, 66, 129
interior 99
lengthened 152, 158-159
stowage compartments 16
surplus 110
Hydraulic systems 115-116, 134, 136, 139-143

Icken, Col Ing 23
Indian Army 132
Infrared decoy system (LIR 30) 56-57, 61, 63
Institut Franco-Allemand de St Louis 89
Instrumentation 125, 128
Iraqi Army 126-127, 161, 164, 166-170
Israeli Army 41

Krauss-Maffei/Krauss-Maffei-Wegman 132, 174
Kuwaiti Army 128

Liberation of France 35
Lights
hull-mounted rear red lamps 145
infrared driving lights 53
infrared searchlight 29, 93
interior 106-107
running 106
searchlights 28, 53-54, 67, 90-92, 113, 167
Losses in combat 161, 169-170

Maintenance and servicing 99-107, 115, 166
inspections 100-103
oil leaks and levels 101-102, 105
powerpack exchange 105-106, 110-112
shared tasks 103-104
turret mechanics 100
weapons 103-104
MAS 66
Messmer, Pierre 27, 150
Missiles 171
anti-tank 41, 171
Crotale 135
Euromissile 132
Honest John nuclear 114, 116
Mistral system 134
Pluton nuclear 114-119
Roland 1 and 2 132-133, 137
SABA 137
Shahine SA-10 (R460) 134-136, 159
Missile-carrying trucks 116-118
Missile control and guidance systems 118-119, 132-137
command microwave radio link 133
radar target acquisition and tracking 132-138
Molinié, Joseph 6, 150
Musée des Blindés de Saumur 11, 71, 101, 107, 143, 171, 175

Names
Cne Rouvillois 5-6
Herbsheim 36
Herbsheim 1945 129
Souville 124
NATO 10, 14, 27, 41, 43, 68, 95, 121-122, 127, 131, 138, 150, 175
NBC air filtration 13, 15, 53-54, 77-81, 111
Netherlands Army (KL) 23, 127, 150
Numbering 7, 15, 22, 48, 51, 62, 66, 132

Operation Azure 58
Operation Daguet 6, 61, 88, 142-144, 164, 166
Operation Desert Storm 60, 128, 170
Operation Vulcain 127

Panhard AML 164
Performance 6, 24, 122
 breakdowns 24
 cross-country 69, 129, 136-137, 154, 158
 power-to-weight ratio 24
 reliability 69, 168
 speed 37, 122
Poswick, Belgian defence minister 150
Preserved and surviving vehicles 7, 11, 71-72, 101, 107, 143, 147, 170, 174-175, 183
Production 20, 27-34, 49, 51, 53, 62, 86, 121, 125
 deliveries 29-30, 37, 53, 57, 60, 126, 128-129, 134, 151, 154, 157, 159
 figures 27, 30, 46, 51, 53-54, 57, 76, 116, 119, 125-128, 133-134, 142-145, 151, 153, 157-160
 last built 34, 38, 71
 periods 36, 71, 93, 126, 134, 142, 151, 160
 subcontracting 33
 under licence 150-151
 unit cost 40, 119, 125, 131
Production facilities 28
 ARE (*Arsenal de Roanne*) 27-31, 33-34, 47, 51, 54, 59, 71, 110, 119, 125-126, 128, 139, 142, 159-161
 ATS (*Atelier de Tarbes*) 28, 31, 33, 66
 EFAB (*Et. de Fabrication d'Armes de Bourges*)/(f. *Arsenal de Bourges*) 28, 30-31, 48, 51, 84-86, 119, 121, 123, 125-126, 136
 Limoges 28, 60

Qatari Army 169

Rail transport 113, 121-123, 126, 166
Reference vehicles 143
Remote-controlled mine-clearing vehicles 144, 146-147
Retirement and replacement 7-8, 45, 50, 58, 61, 130, 134, 157-158, 170
 AU F1 128
 last regiment equipped with AMX30B2 61
 scrapping 157
 storage 8, 61, 63
Roles
 FORAD 8, 58, 61, 63, 96, 114
 OPFOR 40
 UNIFIL 130
 UN peacekeeping 160
Royal Saudi Arabian Army 126, 137, 144, 159-160, 167, 169
 Saudi National Guard 161, 170
Running gear 70
 brake system 105
 road wheels 21-22, 25, 37, 70-72, 102, 121

SAMM 136
Second World War 7-9, 14, 84, 113
Section Technique de l'Armée (STA) 29
Serbian Army 127
Service de Poudres research establishment, Begerac 123
Sights, optics and vision devices 5, 7, 17, 30, 42, 49-50, 83-97, 103, 124, 128, 155, 158, 169
 AMX30B 90-95
 binocular 91, 94, 103
 episcope 42, 93, 103, 124
 image-intensification cameras 53-54, 59, 96-97
 infrared 43, 45, 91, 94, 103
 night vision 50, 91, 95, 167
 periscopes 16-17, 20, 56, 60, 78-79, 90, 93-94, 107, 124
 rangefinders 14-15, 17, 20, 23, 41-43, 45-46, 49, 51, 53, 55, 90-91, 94, 97, 103, 158
 telescopic 49, 92, 96, 101, 128
 television cameras 50-51, 53
 thermal cameras 7, 55, 61-63, 96, 155, 164
Siren 106
SOFMA 150
Spanish Army 151-155
Spanish Foreign Legion 151
Spares 8, 63, 159, 161, 170
Specifications 66, 178-179
 height 122
 weight 7, 18, 23, 26, 114-115, 122
STAT 48, 51, 54-57, 61, 63, 132-133, 146, 173-174

Steering 105-107
Suspension 18, 22, 28, 37, 47-48, 51, 54, 56, 69-72, 105, 121, 125, 159
 hypropneumatic 158-159
 shock absorbers 21-22, 70, 105-106
 swing arms 106
 torsion bar 22 69, 71, 125, 151, 158
 upgrades 154
 Vickers-type 16, 69
Swiss Army 124

Tanks (Brazilian)
 ENGESA Osorio 172
Tanks (British)
 Centurion Mks 5 and 7 150
 Challenger 1 172
 Chieftain 8, 138
 Vickers Mk 7 172
 Vickers Valiant 172
Tanks (collaborative)
 FINABEL 3A5 proposal 9, 14-16, 30, 34, 68, 175
Tanks (French) – see also AMX30
 AMX10P 40, 86, 169; AMX10RC 46, 138; AMX10 RCR 167
 AMX13 10-11, 14, 16, 36, 42, 92, 104, 124, 132, 136, 144, 150, 158
 AMX13 Bitube 30mm DCA 136-137
 AMX40 51, 55, 71, 128, 130, 172-175
 AMX50 M4 9-11, 14, 92
 AMX56 Leclerc 7-8, 58-59, 61, 63, 72, 128, 130, 172; EAU 160, 174-175
 EPG project 142
 ARL44 9
 Char BC 11, 15
 CL 40t 10-11
 ECP programme 172-173
 Renault Char B1 (bis) 9
Tanks (German) 84
 Leopard 1 8, 23, 30, 38, 150-151, 174-175
 Leopard 2 49; 2A4 155, 172, 175; 2F 155
 Leopard GCT 131, 132
 Marder *Schutzenpanzer* IFV 133, 154
 Panther 10, 17, 113
 Standardpanzer prototypes 18, 23, 27, 70
 Tiger 9, 16-17
Tanks (Indian)
 Arjun 132
Tanks (Iraqi) 168
Tanks (Soviet)
 T-54 67
 T-55 16, 161, 164
 T-59 128
 T-62 164
 T-64 40, 45
 T-72 45, 164, 166, 172; T-72 M1 132
Tanks (US)
 M1 Abrams 49, 172, 175
 M4 11
 M26 11, 70
 M47 Patton 6, 10-11, 14, 19, 24, 35-37, 57, 69-70, 92, 110, 151-152
 M48 153; M48A1 152-153
 M60A1 8, 26, 40, 152, 159; M60A3TTS 154
 RISE Passive 45
 M74 tank recovery vehicles 110
 M109 self-propelled gun 131
Theatres of operation
 Afghanistan 8
 Balkans 60
 Bosnia 127, 160
 French borders 24
 Iran 127
 Iraq 8
 Kosovo 63
 Kuwait 7, 144, 160-161, 164, 166-170
 Saudi Arabia 161, 169
 Serbia 128
 Spanish Sahara 151
 Venezuelan-Colombian border 158
 West German border 7
 Yemen 160-161
Thomson-CSF 119, 132-133, 135, 138
Tracks 24, 48, 56, 63, 70-72, 125, 152, 158
 guards 17, 159-161
 idlers and sprockets 71, 102
 removing and replacing 72-73

return rollers 18, 70-72
 sand skirts 166, 169
 tension 105
Training and testing grounds 43-44
 ARE 34
 Bergen-Hohne 22
 Bourges 18-19, 23, 29, 79, 115, 123, 125
 Canjuers 115
 Carplagne driver's training centre 37
 Mailly 8, 20-25, 27, 35, 37, 40, 58, 96, 114, 150, 173
 Meppen 18, 20, 23
 Mourmelon 22, 51, 164, 166, 169, 175
 Munsingen 22-23
 Satory 10, 17-20, 23, 25, 79
  arms exhibition 1967 35; 1979, 46
 Sissonne 7-8, 58
 Valdahon 173
Training manuals 27, 66, 123
Transmission 37, 41-42, 51, 68-73, 104, 130, 152
 Allison CD 850 automatic 151-154, 158
 ESM500 172
 failures 37, 69, 104, 151-152, 154
 final drive 105-106
 gearboxes 23, 28, 37, 48, 104-105, 152, 154
 Minerva ENC 200 automatic 42, 45, 48, 51, 54, 57, 60, 142, 152
 powertrains 31, 41, 47, 152, 159
 Renk ZF LSG 173, 175
Treaty on Conventional Armed Forces in Europe 1990 154
Trials and testing 15-24, 29, 36, 47-51, 56, 61, 149, 153, 155
 AMX30D 110
 AMX30P 115
 cold weather 24
 comparative 16, 18, 24, 27
 demining 146
 desert 161
 EBG 140
 evaluations 16, 22, 132, 137, 173-174
 firing 57
 fording 123, 147
 gunnery 18-19, 87
 platoon 22
 suspension 22
 tactical 23
 technical 19-20, 23
 tripartite 16, 24
Turrets 7, 14, 16-17, 28, 31, 33, 37, 41, 47, 51, 54, 59, 66-67, 91, 99
 bustle 68, 76, 79-81, 97, 102
 CAFL 20
 EBG 141
 GCT 33, 119, 121-122, 124-126, 128-132, 138
 gun cradles 21
 hoisting off 100
 hydraulic systems 101-102
 interior 86, 88
 lock 78
 mantlet 67, 156
 modernisation 154
 Roland 132-134
 roof gun clamp 85
 Sabre system 138
 SAMM S401A 136-137
 traverse 66-67, 90, 96, 101-102, 134, 137
 TG230A 137-130
 T105M 80, 94
 T142 171
 welded 46, 48

US Army Aberdeen Proving Ground 11
US MAP funding 10-11, 14, 36, 110, 136
US Marine Corps 161, 169

VAB 4x4 APC 134, 147, 165, 167
Venezuelan Army 152, 158

Water-fording 19, 25, 44, 77-79, 112, 123, 125, 146, 157
 combat divers 78, 112
 ditching 45
 preparations 77
 snorkels 44, 56, 77-79
 training tower 78-79
Waterproofing equipment 25, 53, 78-79, 147
West Germany 14-15, 22, 27, 132, 150

Yom Kippur War 1973 41, 90